Landslides - Investigation and Monitoring

Edited by Ram Ray and Maurizio Lazzari

Published in London, United Kingdom

IntechOpen

Supporting open minds since 2005

Landslides - Investigation and Monitoring
http://dx.doi.org/10.5772/intechopen.78130
Edited by Ram Ray and Maurizio Lazzari

Contributors
Yi-Min Huang, Ihui Chen, Miaubin Su, Sheichen Ho, Yushu Lin, Junyang Chen, Jónas Elíasson, Thorsteinn Saemundsson, René Vázquez-Jiménez, Rocío N. Ramos-Bernal, Sulpicio Sánchez Tizapa, Roberto Arroyo Matus, Herman Peiffer, Fabio Luino, Laura Turconi, Ram Ray, Maurizio Lazzari, Tolulope Olutimehin, Marco Piccarreta, Salvatore Manfreda, Tsu-Chiang Lei, Bing-Jean Lee, Meng-Hsun Hsieh

First published in London, United Kingdom, 2020 by IntechOpen
IntechOpen is the global imprint of INTECHOPEN LIMITED, registered in England and Wales, registration number: 11086078, 5 Princes Gate Court, London, SW7 2QJ, United Kingdom
Printed in Croatia

British Library Cataloguing-in-Publication Data
A catalogue record for this book is available from the British Library

Additional hard and PDF copies can be obtained from orders@intechopen.com

Landslides - Investigation and Monitoring
Edited by Ram Ray and Maurizio Lazzari
p. cm.
Print ISBN 978-1-78985-823-5
Online ISBN 978-1-78985-824-2
eBook (PDF) ISBN 978-1-83968-285-8

For EU product safety concerns:
IN TECH d.o.o., Prolaz Marije Krucifikse Kozulić 3, 51000 Rijeka, Croatia, info@intechopen.com or visit our website at intechopen.com.

Meet the editors

Dr. Ram L Ray is an associate professor at Prairie View A&M University, USA. He also worked as a research scientist at Prairie View A&M University for six years. Dr. Ray obtained his PhD in Civil Engineering from the University of New Hampshire in 2009, his MS in Soil Science from the Vrije Universiteit Brussel and Ghent University, Belgium in 2004, and his BS in Civil Engineering from Tribhuvan University, Kathmandu, Nepal in 1993. He was a postdoctoral researcher at the University of California (UC) and the San Diego State University and UC San Diego, USA, from 2009 to 2013. From 1996 to 2002, Dr. Ray worked as a civil infrastructure engineer for the Government of Nepal, and from 1993 to 1996 he worked as a civil engineer at the TAEC Consult P. Ltd. He is a member of the American Society of Civil Engineering (ASCE), Soil Science Society of America (SSSA), and the American Geophysical Union (AGU). Dr. Ray's research interests include natural hazard (landslide and flood) and risk analysis using GIS/remote sensing and spatial statistical analysis, fluvial geomorphology, flood risk analysis, flood hazards, soil water dynamics, water resource management, hydrologic modeling, remote sensing, climate change, carbon sequestration, and precision agriculture. He has several peer-reviewed publications and book chapters on landslide hazards, flood prediction, soil moisture dynamics, carbon sequestration, drought, precision agriculture, climate, and land use/land cover change studies using remotely sensed data, model, and field experiments. He is editor and editorial board member of several journals including *Remote Sensing in Earth System Sciences*, *Remote Sensing*, *Agronomy*, *Civil Engineering*, *Constructions*, and others.

Dr. Maurizio Lazzari has a PhD in Earth Science and is a researcher at the Italian National Research Council, Institute of Cultural Heritage Sciences. Since 2001, Dr. Lazzari has been Professor of Pedology at the University of Basilicata (Italy) and a geoarchaeologist at the University of Salento (Italy). His research activities are focused on natural and anthropic hazard and risk factors, aimed at safeguarding and conserving settlements and the historical–monumental heritage of the Mediterranean, with particular attention to landslide processes, susceptibility maps, monitoring, and modeling. Since 2004, he has been working as a scientific coordinator for several national research projects studying landslides and triggering factors, natural and anthropogenic risks, geological and geomorphological mapping, soil erosion, preservation of historical and archaeological sites, enhancement of degraded areas, geo-touristic use, and the protection of the landscapes. He is the author of about 150 scientific publications in national and international journals, monographs, book chapters, and conference proceedings concerning applied geology, geomorphology, dynamics of artificial reservoirs, soil erosion, landslides, geoarchaeology, hydrogeological instability, natural hazards, monitoring, cultural landscape, UNESCO Heritage, geoarchaeology, and geo-tourism.

Contents

Preface

This book was born from the desire to offer to a broad audience of interested readers an overview of cognitive approaches to the study of landslides and the control of their evolution, through innovative integrated monitoring techniques. This subject is very topical, considering that landslides are one of the most destructive natural hazards, as they cause significant threats to lives, properties, and natural environments throughout the world, especially in mountainous regions.

This book provides the reader with a comprehensive overview of the current state of the art in landslide disasters, featuring an easy-to-follow format that focuses on the most important evidence-based developments in this critically important area.

Several authors, belonging to research and public institutions from different parts of the world, have collaborated in the technical discussion of this topic, reporting their experience and present advances in the critical research areas. The book contains nine chapters that cover important research aspects inherent to mapping, susceptibility, hazard and risk, monitoring, and modeling.

The introductory chapter, **Chapter 1**, prepared by the editors, consists of an overview of “Landslide Hazards” that includes overarching concepts of importance to the study of landslides, including global landslide information, factors affecting landslides, and recent trends and techniques used to investigate landslides. **Chapter 2**, prepared by Ram L. Ray, Maurizio Lazzari, and Tolulope Olutimehin, discusses remote sensing approaches and techniques used to study landslides and explores the possibilities of potential remote sensing tools that can effectively be used in landslide studies in the future. **Chapter 3**, prepared by Fabio Luino, and Laura Turconi, explains the unique translational rock-block slides located in Tertiary Flyschid Complexes of the Southern Piedmont Region (North-West Italy).

Chapter 4, prepared by Rocío N. Ramos-Bernal, René Vázquez-Jiménez, Sulpicio Sánchez Tizapa, and Roberto Arroyo Matus, characterizes susceptible landslide zones by using the accumulated index. This study helps to locate areas vulnerable to landslides and integrate disaster management or prevention plans. **Chapter 5**, prepared by Maurizio Lazzari, Marco Piccarreta, Ram L. Ray, and Salvatore Manfreda, explains the role of antecedent soil moisture on critical rainfall intensity-duration thresholds to evaluate the possibility of modifying or improving traditional approaches.

Chapter 6, prepared by Miau-Bin Su, I-Hui Chen, Shei-Chen Ho, Yu-Shu Lin, and Jun-Yang Chen, presents time-domain reflectometry (TDR) technology for a long-term landslide monitoring system to explore rock deformation mechanisms and detect the location and magnitude of slope movement. **Chapter 7**, prepared by Jónas Elíasson and Þorsteinn Sæmundsson, presents the use of the translatory wave model to study landslides. This chapter also includes three case studies to study submarine slides (Japan), rock avalanche (SE Iceland), and debris slides (N Iceland) using the translatory wave model.

Chapter 8, prepared by Herman Peiffer, discusses the techniques that can be used to evaluate and remediate the slope stability in overconsolidated Boom Clay in Kruibeke, Belgium, the same origin as London Clay. Finally, **Chapter 9**, prepared by Yi-Min Huang, Tsu-Chiang Lei, Bing-Jean Lee, and Meng-Hsun Hsieh, explains the use of the fragility curve model to evaluate potential landslides.

Overall, this book significantly contributes to understanding the potential causes of landslides, advancing methods in monitoring, investigating, mapping, and mitigating.

The editors wish to express their thanks to all the participants in this book for their valuable contributions, and Ms. Kristina Kardum Cvitan (Author Service Manager) for her assistance in finalizing the work. We also acknowledge the staff at IntechOpen for their help in publishing this book and others.

Ram Ray
Prairie View A&M University,
Prairie View, Texas, USA

Maurizio Lazzari
CNR-ISPC,
Potenza, Tito Scalo, Italy

Chapter 1

Introductory Chapter: Importance of Investigating Landslide Hazards

Ram L. Ray and Maurizio Lazzari

1. Introduction

Natural disasters, like earthquakes, volcanic eruptions, intense rainfall, and anthropogenic factors, such as deforestation, contribute to slope failure, either by decreasing resisting forces or increasing driving forces of the soil mass [1]. Landslides are strongly related to steepness of the slope, soil moisture/water content of the soil layer, climate factors that increase the water content of the soil, and other anthropogenic factors and can be triggered by earthquakes, volcanoes, and floods. However, most of the slope failures are proceeded by intense rainfall and wet antecedent soil moisture conditions [2–4]. Often aggravated by rapid and uncontrolled development, landslides, either large or small, which happen every year in mountainous regions around the world [5, 6].

To better understand and manage the potential landslides, it is essential to know the location and size of potential slope failures [7]. However, it is a difficult task to predict precise size and location of the possible landslides. Since slope failure is a complex phenomenon, it requires an in-depth understanding of slope failure mechanisms and monitoring techniques.

It is necessary to collect/obtain high-resolution spatial information of the soil layer, topography, hydrologic conditions, geotechnical characteristics, and land use/land cover types to investigate landslide, including mapping, detection, monitoring, analysis, prediction, and others. Since slope failures commonly occur in the hilly region, especially in steep terrain, so it is rather challenging to obtain high-resolution data in conducting landslide studies [8, 9].

It is always recommended to obtain in-situ measurements for an accurate landslide study. However, such in-situ measurements are time-consuming and require complex data collection efforts even on local scales [2, 6]. Recently, remote sensing data and spatial analysis tools are widely used in landslide studies, including landslide detection, assessment, hazard, mapping, and inventories [10–12]. Remote sensing data makes it possible to conduct landslide studies, not only at inaccessible terrain but also at regional to global scale, which otherwise is not possible using in situ measurements.

2. Importance of information of historic landslides

Data archives of national projects, besides constitute a sort of historical encyclopedia, they also represent a potential operational support tool useful and functional for planning and managing territorial and mitigation policies for landslide risks.

The awareness that territorial planning and emergency plans can provide significant information from historical data series on localities and areas previously affected by hydrogeological disasters have recently stimulated the international

 IntechOpen

scientific community to systematically collect data on landslides and floods [13]. Landslides, statistically, represent, after earthquakes and other external driving forces of natural disasters, which cause the most significant number of victims and damages built-up areas, infrastructures, environmental, historical, and cultural assets.

In particular, those which cause the most damage are fast-moving landslides (rockfalls, rapid mudflows, and debris flows), as well as those which involve large volumes of rocks or soils. Several landslide investigation projects, economically supported by public agencies, tended to focus mainly on phenomena that caused significant or evident damages (preferably occurring in urban areas or correspondence with linear infrastructures). They often neglected some landslides, even a substantial entity, which had not affected or interacted with built-up areas, communication routes, or infrastructures.

Historical memory helps play a fundamental and decision-making role that can be recovered and used through integrated methodological approaches. The collection of historical data aimed at knowledge and mapping of the instabilities allows us to complete and improve the information obtained with normal geological-geomorphological analysis, better defining fundamental aspects for assessments of danger and vulnerability of the region.

3. Global landslide information

Knowledge of past landslides of a region, in terms of the occurrence of the event, controlling factors, and trigger conditions, are the main needed factors to evaluate spatial and temporal probabilistic hazard [14]. Therefore, the use of published and unpublished historical sources (federal, municipal, provincial, and state archives, publications, newspapers, press reviews, technical reports, photographs, and videos) can provide essential frameworks to understand the impact of hazard events over time in the areas monitored and evaluated during the new surveys.

Historical information can be grouped into four categories:

1. Direct recorded changes or natural events, such as droughts, floods, landslides, and erosion rates;

2. Indirect data used to determine causes or explain patterns, such as historical rainfall series;

3. Other data that provides additional information, such as geological maps;

4. phenomenological data that may change with time (e.g., the response of aquifers to wet seasons).

The types of information available from local archive data of ancient inscriptions, annals, historical chronicles, private funds, ecclesiastical funds, newspapers, iconographies, magazines, monographs, old postcards, cartographies, and videos (**Figure 1**) are broad and may differ from place to place.

The data obtainable from these sources is extensive and vital but requires control and validation of the news and associated information, such as geographical (e.g., location, municipality, street) and temporal ones (exact day, time, etc.) or climatic conditions (rainfalls mm/h, local measurement stations, etc.).

The historical-environmental analysis uses documentary information to reconstruct impact scenarios of natural events of the past on the environmental and

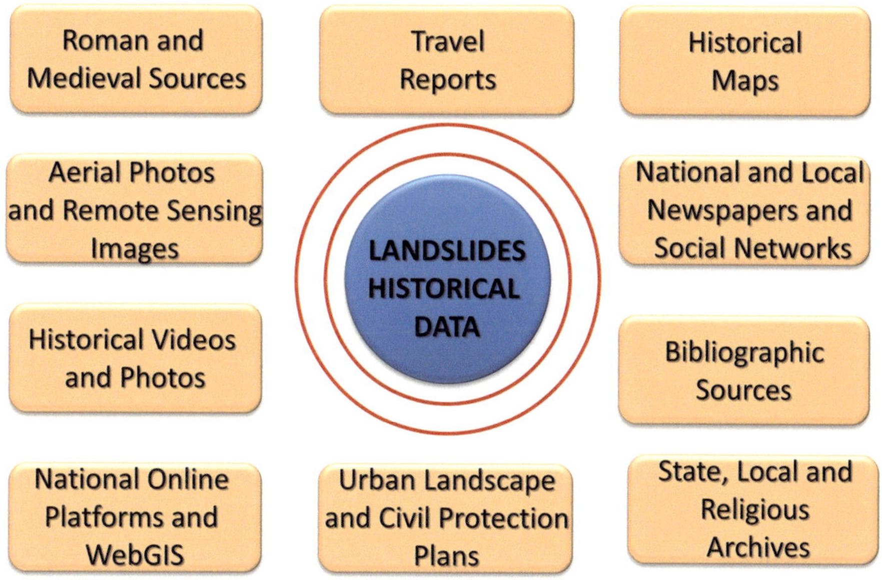

Figure 1.
Synthesis of the remarkable diversity of historical documents potentially available to carry out geologic hazard analysis by integrating the present-day scenarios and data.

anthropic context of the time [15]. A fundamental method of interpreting historical information is to bring it back to its cultural, political, and economic context. The degree of reliability of data is, in fact, directly proportional to the knowledge, more or less in-depth, of this context.

4. Factors affecting landslides

There are numerous internal and external forces, natural and anthropogenic factors that trigger landslides [16–19]. The primary natural and anthropogenic factors that triggered landslides are outlined below.

4.1 Natural factors

Topography - Steep terrain
Gravity - slope failure due to gravitational forces
Geological- weathered/sheared/susceptible materials
Morphological- fluvial/glacial/wind erosion of slope toe, tectonic, volcanism
Heavy and Prolonged Rainfall -
Hydrological- rise in the groundwater table, flooding, rapid snowmelt
Earthquakes - ground vibrations created during Earthquakes
Climate – cause series of natural disasters (e.g., drought, floods, storm, hurricanes, etc.)

4.2 Anthropogenic factors

Deforestation – enhance soil erosion and surface runoff
Overloading Slopes – increase surcharge load
Mining and Quarrying Activities – increase instability
Construction – increase instability

The driving forces of landslides are physical/geological, morphological, and human in nature. Globally, the prominent causes of landslides are geological in nature and rainfall-induced. Landslides occurring majorly in mountainous and coastal terrains have also occurred in plains, which can be such as failures of the roadway and building, in addition to quarries and open-pit mines. These are sometimes the resultant effects of heavy rainfalls, volcanic eruptions, earthquakes, and droughts. Some areas are susceptible to landslides due to human activities affecting vegetation and topography. This also happens in places where wildfires occur.

Landslides are unpredictable. For example, landslides can occur due to human activities or non-human activities that affect slope stability [20]. Geological factors may account for 43% of landslides. This includes the impact of gravity on the topography of sloped areas, water pressure, weak soil formation, etc. The morphological causes, such as volcanic pressure, underground erosion, climate factors, vegetation elimination, crest accumulation. Human activities, such as excavation, irrigation, mining, deforestation, and slope encroachment, also enhance slope instability [21, 22]).

Kazmi et al. [21] investigated 11 landslide events in Malaysia and summarized water movements, weak safety management, heavy downpours, inadequate slope protection, damaged drainage, flawed design, and construction were some of the underlying factors that triggered landslides. Singh and Singh [23] found urbanization as a contributing factor for the risk of landslides and hazards. They discussed one of the most famous landslide events in the history known as Frank Slide. The Frank Slide of Turtle Mountain of Canada occurred in 1903 and generated 82 million tons of limestone. The primary cause of this landslide was the geology of the mountain because weak rock and stones were covered by limestone rock. Another factor was the weather event prior to the landslide, which was more snowy than usual, which allowed snowmelt and rain to permeate the mountain. The resistance forces of the rocks were more weakened beyond bearable limits. Pal et al. [24] reviewed several other landslide events and found intense rainfall as a common triggering factor for these landslide events. Since the causes of landslides across the world are multi-faceted, more advanced researches are necessary to investigate the triggering factors of landslide events.

5. Landslide investigation: recent trends and techniques

Landslide investigation has been promoted by the International Decade of Natural Disaster Reduction (IDNDR, 1990–2000) proclaimed by the United Nations, when working groups on landslides were established (e.g., International Landslide Research Group (ILRG)). In addition, recent advances in landslide investigations include real-time monitoring, modeling, prediction, and assessment, which are helping communities and end-users to be better prepared to face potential landslide threats [25]. During the past three decades, tremendous developments have been made to investigate landslides. The investigation of Landslides has been carried out using surface and subsurface methods comprising qualitative and quantitative approaches. Although the wide range of applied geophysical techniques were used in landslide investigations, primarily these are grouped into two main classes: remote sensing techniques, which can be used to characterize the Earth's surface, and sub-surface techniques, which characterize geological surface by using non-destructive approaches [26]. In addition, an integrated approach combining satellite, airborne, and ground-based sensing, is widely used to investigate landslides [27, 28]. The wide range of remote sensing data from various optical and microwave synthetic aperture radar (SAR) sensors and the increased temporal and spatial resolution provides new opportunities to investigate landslides at a range of scales [26, 29].

Some other advancements for landslide investigation are the use of in-situ geophysical techniques and electrical resistivity tomography (ERT), which can be used for landslide detection [27]. The interferometric techniques, which include Multi temporary interferometry (MTI) and advanced synthetic aperture radar differential interferometry (DInSAR) techniques, can be used to extract information on ground surface deformations. Also, interferometry SAR (InSAR) data combined with Unmanned Aerial Vehicle (UAV) images and aerial photography can be used to investigate slow-moving landslide [30]. Recently, application of UAV is growing to monitor rapidly occurring landslides and mapping in inaccessible terrains [31, 32].

Machine learning techniques are also getting popular in evaluating and detecting landslides [33, 34]. In addition, the use of artificial intelligence (AI) technique is growing to investigate landslides, such as landslide susceptibility mapping, characterization, and prediction [35].

Currently, remote sensing technologies are used in landslide monitoring, mapping, hazard prediction and assessment, inventory and detection, and other investigations. Some of the key technological advancements for investigating landslides outlined by Petley [26] are as follows:

Digital imaging – Digital photogrammetry to capture image of landslides.

Optical satellite sensors – Many commercial satellites are providing high spatial resolution data to study landslides (e.g., GeoEye-1 (0.4 m); Quickbird (0.6 m)).

Google Earth – An important tool for visualization and analysis.

Radar satellite sensors – Radar techniques allow to accurately determine the small landslide movement.

Terrestrial laser scanners (TLS)- The TLS, a ground-based LiDAR, mainly used to collect field data on steep unstable terrains.

Airborne LiDAR – A little expensive to collect data; however, it is widely used to collect high spatial resolution data to study landslides.

Engineering geophysics – This geophysical technique is widely used and applied to evaluate the sub-surface environment of landslides in 3D and in real-time.

6. Concluding remarks

Landslide, a catastrophic disaster, has been on the rise due to the impact of natural and anthropogenic, such as climate and land use/land cover change, and growing population. Landslides are common around the world, especially in mountainous regions. Since landslides are a severe threat to lives and properties, it is essential to understand the physical processes, causes of landslides, movement characteristics, and potential risk factors. It is also vital to study landslides, which helps understand landslide mapping, prediction, monitoring, and risk assessment to reduce the impact of landslides. However, such in-depth landslide investigations require advanced technologies, robust methods, models, and high resolution spatial data, which includes in-situ and/or remotely sensed measurements globally.

While a high resolution data is required for landslide investigation, the potential landslide area is mostly inaccessible, which limits for in-situ measurements. Although recently, the application of satellite for landslide studies is growing, high spatial and temporal resolution satellite data are still limited on a global scale. Regardless of recent advancements in landslide studies, more research efforts, advanced technologies, and tools, high resolution spatial and temporal data, and effective management, awareness, and policy are needed in landslide research. It will help address the impact of future human activity, climate change, land use/land cover change on landslide hazards from local to the global scale.

Author details

Ram L. Ray[1*] and Maurizio Lazzari[2]

1 College of Agriculture and Human Sciences, Prairie View A&M University, Prairie View, Texas, USA

2 CNR-ISPC, C/da S. Loja, Zona Industriale, 85050 Tito Scalo (PZ), Italy

*Address all correspondence to: raray@pvamu.edu

References

[1] Abramson LW, Lee TS, Sharma S, Boyce GM. Slope stability and stabilization methods. Wiley, New York. 1996. 738 pp.

[2] Ray RL, Jacobs JM, Douglas EM. Modeling regional landslide susceptibility using dynamic soil moisture profiles. J. Mt. Sci. 2018;15(8):1807-1824. DOI: 10.1007/s11629-018-4896-3.

[3] Lazzari M, Piccarreta M, Manfreda S. The role of antecedent soil moisture conditions on rainfall-triggered shallow landslides. Nat. Hazards Earth Syst. Sci. 2018. DOI: 10.5194/nhess-2018-371

[4] Ray RL. Jacobs JM. Relationships among remotely sensed soil moisture, precipitation and landslide events. Nat. Hazards. 2007;43:211-222. DOI: 10.1007/s11069-006-9095-9.

[5] Chuang Y-C, Shiu Y-S. Relationship between landslides and mountain development—Integrating geospatial statistics and a new long-term database. Science of the Total Environment. 2018;622-623:1265-1276. DOI: 10.1016/j.scitotenv.2017.12.039

[6] Ray RL, De Smedt F. Slope stability analysis on a regional scale using GIS: a case study from Dhading, Nepal. Environ Geol. 2009;57:1603-1611. DOI: 10.1007/s00254-008-1435-5.

[7] Davies T. Landslide Hazards, Risks, and Disasters. IN: Shroder, J.F. (editor), Hzards and Disaster Series, Elsevier. 2015. 475 pp.

[8] Singh TN, Gulati A, Dontha L, Bhardwaj V. Evaluating cut slope failure by numerical analysis - a case study. Natural hazards. 2008;47:263. DOI: 10.1007/s11069-008-9219-5

[9] Ray RL, Jacobs JM, Ballestero TP. Regional landslide susceptibility: spatiotemporal variations under dynamic soil moisture conditions. Nat. Hazards. 2011;59:1317-1337. DOI 10.1007/s11069-011-9834-4.

[10] Ray RL, Jacobs JM, Cosh MH. Landslide susceptibility mapping using downscaled AMSR-E soil moisture: A case study from Cleveland Corral, California, US. Remote Sensing of Environment. 2010;114:2624-2636. DOI: 10.1016/j.rse.2010.05.033.

[11] Pradhan B, Singh RP, Buchroithner MF. Estimation of stress and its use in evaluation of landslide prone regions using remote sensing data. Advances in Space Research. 2006;37:698-709. DOI: 10.1016/j.asr.2005.03.137.

[12] van Westen CJ. GIS in landslide hazard zonation: A review, with examples from the Andes of Colombia. In: M. F. Price, D. I. Heywood (eds). Mountain environments and geographic information systems. Taylor and Francis Publishers. 1994. p. 135-165.

[13] Tropeano D, Turconi L. Using Historical Documents for Landslide, Debris Flow and Stream Flood Prevention. Applications in Northern Italy. Natural Hazards. 2004;31(3):663-679. DOI: 10.1023/B:NHAZ.0000024897.71471.f2

[14] Ibsen ML, Brunsden D. The nature, use and problems of historical archives for the temporal occurrence of landslides, with specific references to the south coast of Britain, Ventnor, Isle of Wight. Geomorphology. 1996;15:241-28. DOI: 10.1016/0169-555X(95)00073-E

[15] Lazzari M, Geraldi E, Lapenna V, Loperte A. Natural hazards vs human impact: an integrated methodological approach in geomorphological risk assessing on Tursi historical

site, southern Italy. Landslides. 2006;3(4):275-287, Springer-Verlag.

[16] Gue SS, Tan YC. Landslides: Abuses of the Prescriptive Method, International Conference on Slope, Kuala Lumpur, Malaysia: 2006. p. 34-42.

[17] Jamaludin S, Ali F. An overview of some empirical correlations between rainfall and shallow landslides and their applications in Malaysia. Electron J Geotech Eng. 2011;16:1429-1440.

[18] Lee S, Pradhan B. Landslide hazard mapping at Selangor, Malaysia using frequency ratio and logistic regression models. Landslides. 2007;4(1):33-41. DOI: 10.1007/s10346-006-0047-y.

[19] Azmi AS, Salleh WA, Nawawi AH. Cognitive behaviour of residents toward living in landslide prone area: Ulu Klang. Procedia - Social and Behavioral Sciences. 2013;101:379-393. DOI: 10.1016/j.sbspro.2013.07.212.

[20] Walker LR, Shiels AB. Physical causes and consequences. In: Walker LR, Shiels AB (eds) Landslide Ecology. Cambridge University Press; New York, NY. 2013. 46-82.

[21] Kazmi D, Qasim S, Harahap I, Baharom S, Imran M, Moin S. A Study on the Contributing Factors of Major Landslides in Malaysia. Civil Engineering journal. 2016;2(12):669-678. DOI: 10.28991/cej-2016-00000066.

[22] Dlamini W. Analysis of deforestation patterns and drivers in Swaziland using efficient Bayesian multivariate classifiers. Model. Earth Syst. Environ. 2016;2:173. DOI: 10.1007/s40808-016-0231-6.

[23] Singh K, Singh MP. Causes and remedial measures for rockfall and landslides in Naini lake basin: Uttarakhand, India. Environment Conservation Journal. 2020;21(1&2):95-102. DOI: 10.36953/ECJ.2020.211211

[24] Pal R, Biswas SS, Mondal B, Pramanik MK. Landslides and Floods in the Tista Basin (Darjeeling and Jalpaiguri Districts): Historical Evidence, Causes and Consequences. J. Ind. Geophys. Union. 2016;20(2):66-72.

[25] Gutierrez F, Soldati M, Audemard F, Balteanu D. Recent advances in landslide investigation: issues and perspectives. Geomorphology. 2010;124(3-4):95-102. DOI: 10.1016/j.geomorph.2010.10.020

[26] Petley DN. Remote sensing techniques and landslides. In: Clague J, Stead D(eds) Landslides types, mechanisms and modeling. Cambridge University Press; New York, NY. 2012. p. 159-71.

[27] Perrone A, Lapenna V, Piscitelli S. Electrical resistivity tomography technique for landslide investigation: A review. Earth-Science Reviews. 2014;135:65-82. DOI: 10.1016/j.earscirev.2014.04.002

[28] Scaioni M, Longoni L, Melillo V, Papini M. Remote sensing for landslide investigations: An overview of recent achievements and perspectives. Remote Sens. 2014; 6:9600-9652. DOI: 10.3390/rs60x000x.

[29] Hölbling D, Friedl B, Dittrich J, Cigna F, Pedersen G. Combined interpretation of optical and SAR data for landslide mapping. In: Advances in Landslide Research, Proceedings of the 3rd Regional Symposium on Landslides the Adriatic-Balkan Region, Ljubljana, Slovenia, 11-13 October 2017; Jemec Auflic, M., Mikos, M.,Verbovsek, T., Eds.; Geological Survey of Slovenia: Ljubljana, Slovenia, 2018; pp. 11-13.

[30] Ekera R, Aydın A. Long-term retrospective investigation of a large, deep-seated, and slowmoving landslide using InSAR time series, historical aerial photographs, and UAV data: The case of Devrek landslide (NW Turkey). Catena.

2020;196:104895. DOI: 10.1016/j.catena.2020.104895

[31] Lindner G, Schraml K, Mansberger R, Hubl J. UAV monitoring and documentation of a large landslide. Applied Geomatics. 2015;8(1). DOI: 10.1007/s12518-015-0165-0

[32] Yaprak S, Yildirim O, Susan T, Inyurt S, Oguz I. The role of unmanned aerial vehicles (UAVs) in monitoring rapidly occurring landslides. Nat. hazards Earth Syst. Sci. Discuss. 2018. DOI: 10.5194/nhess-2018-13

[33] Merghadi A, Yunus AP, Dou J, Whiteley J, ThaiPham et al. Machine learning methods for landslide susceptibility studies: A comparative overview of algorithm performance. Earth-Science reviews 2020;207:103225. DOI: 10.1016/j.earscirev.2020.103225

[34] Wang H, Zhang L, Yin K, Luo H, Li J. Landslide identification using machine learning. Geoscience Frontiers. 2020. DOI: 10.1016/j.gsf.2020.02.012

[35] Chen W, Shirzadi A, Shahabi H, Ahmad BB, Zhang et al. A novel hybrid artificial intelligence approachbased on the rotation forest ensemble and naïveBayes tree classifiers for a landslide susceptibility assessment in Langao County, China. Geomatics, Natural Hazards and Risk. 2017;8(2):1955-1977. DOI: 10.1080/19475705.2017.1401560

Chapter 2

Remote Sensing Approaches and Related Techniques to Map and Study Landslides

Ram L. Ray, Maurizio Lazzari and Tolulope Olutimehin

Abstract

Landslide is one of the costliest and fatal geological hazards, threatening and influencing the socioeconomic conditions in many countries globally. Remote sensing approaches are widely used in landslide studies. Landslide threats can also be investigated through slope stability model, susceptibility mapping, hazard assessment, risk analysis, and other methods. Although it is possible to conduct landslide studies using in-situ observation, it is time-consuming, expensive, and sometimes challenging to collect data at inaccessible terrains. Remote sensing data can be used in landslide monitoring, mapping, hazard prediction and assessment, and other investigations. The primary goal of this chapter is to review the existing remote sensing approaches and techniques used to study landslides and explore the possibilities of potential remote sensing tools that can effectively be used in landslide studies in the future. This chapter also provides critical and comprehensive reviews of landslide studies focus¬ing on the role played by remote sensing data and approaches in landslide hazard assessment. Further, the reviews discuss the application of remotely sensed products for landslide detection, mapping, prediction, and evaluation around the world. This systematic review may contribute to better understanding the extensive use of remotely sensed data and spatial analysis techniques to conduct landslide studies at a range of scales.

Keywords: remote sensing, landslide detection, landslide mapping, landslide inventory, natural hazards, susceptibility, assessment

1. Introduction

Landslides are natural hazards that have a significant impact globally [1, 2]. In comparison to other natural hazards, landslides are one of the costliest and fatal geological hazards, threatening and influencing the socioeconomic conditions of many countries throughout the world [3–5]. A landslide can be triggered by various natural phenomena (e.g., earthquakes, heavy rainfall, tsunami, and flood) and human disturbances (e.g., deforestation, infrastructure developments by cutting slopes, and presence of historical underground cavities) [6–8]. A landslide occurs when the soil layers of the slope get detached either from saturation due to extreme rainfall events or from external forces (e.g., earthquakes) and move downhill causing loss of life, properties, environments, and economic damage. For example, in the U.S. alone, landslides cause approximately $3.5 billion in damage and kill between 25 to 50 people each year [9]. Also, in 2014 alone, Nepal had a landslide where livestock loss was many times larger than human loss and infrastructural damage was more expensive in comparison to the economy of the country [10]. Slope failures also cause major sedimentation into streams and lakes, which further represents a major cause of flooding [5, 11].

Landslides are common primarily in the mountainous regions if a slope is unstable or becomes unstable due to external driving forces [10, 12]. Landslide hazard can be classified into high, moderate, and low based on the volume, duration, possible effect in terms of distance, area, and speed at which the slope fails. Since landslides can adversely affect human lives and property, it is essential to monitor, detect, map, and conduct hazard analysis in order to reduce the impact of their hazard and save human lives, property, and environment around the globe. Landslide susceptibility maps can be developed for landslide-prone regions by combining all of the potential predisposing factors, which cause a landslide.

Landslide susceptibility depends on the local terrain, land use, and climatic conditions, which require spatial information [13]. A landslide susceptibility zone includes information of past landslide inventory with an evaluation of future landslide-prone areas, but it does not include assessment of the frequency of landslide occurrence [14]. Also, the temporal probability of a landslide is not included in susceptibility models [15]. High-quality landslide inventory maps can be developed using in situ measurements and field surveying [16, 17]. However, in situ measurements and field surveying are time consuming, expensive, and difficult for local to global scales. On the other hand, landslide susceptibility and inventory maps, as well as landslide hazard analysis, can be possible using remote sensing techniques and data, such as aerial surveys, unmanned aerial vehicles (UAV), light detection and ranging (LiDAR), and satellite imagery [16, 18].

Although remote sensing is continuously used for landslide detection/mapping and monitoring, it is generally considered to have a medium effectiveness/reliability for landslide studies because satellite data are available relatively at coarse resolutions [3]. On the other hand, hazard assessment requires high resolution data to define the spatial distribution of landslides and their state of activity both on a local scale and from studies from regional/global scales [10]. In addition, remotely sensed data are cost-effective because most of the global satellite products are freely available and can cover rugged/complex terrains, which is otherwise not possible to assess with in situ measurements [19]. Even in the late 1990s, stereoscopic air-photo interpretation was the most used remote sensing tool applied to the mapping and monitoring of landslides [20]. Many studies have been carried out on landslide hazard evaluation using geographic information systems (GIS) and geoinformation-related techniques [4]. Recently, GIS and remote sensing tools have become powerful tools for integrating spatial data to conduct landslide studies [10, 21].

Remotely sensed data and techniques are widely used in landslide studies, including landslide inventory/detection, monitoring, and mapping, and hazard analysis (e.g., [16, 22, 23]). Timely and high-quality information derived from space-borne observations helps in managing natural and human-made disasters [1]. Accordingly, landslide risk mapping and management can help reduce disaster risk [10]. Similarly, early landslide predictions and warnings are important in curtailing landslide hazards [4, 24]. Landslide vulnerability assessment is used in identifying what elements are at risk and why; such information helps in landslide disaster mitigation measures [10, 21].

The use of remote sensing data for landslide studies, whether air, satellite, or ground-based measurements, is mainly classified into three main categories: (a) detection and identification, (b) monitoring, and (c) spatial analysis and hazard prediction [20]. However, we have reviewed landslide studies using multiple categories as they appear in the literature.

This chapter aims to provide a critical and comprehensive review of recent landslide studies conducted using remotely sensed data and techniques. It includes an overview of landslide inventory/detection, mapping and monitoring, susceptibility, and hazard analysis using remotely sensed data at a range of scales. It helps to understand the potential benefits of conducting landslide studies using satellite data to save human lives, properties, and environments around the globe.

2. Landslide inventory/detection

Landslides, influenced by several preparatory and triggering factors, are naturally hazardous events causing loss of lives and properties and environmental degradation [25]. Landslide-triggering factors such as intense or prolonged rainfall or rapid snowmelt, earthquakes, or volcanic eruptions are enhanced by human-induced triggering factors such as deforestation, mining, and cutting slopes for road development [8, 26, 27]. Landslides include various movements like flowing, sliding, toppling, or falling, and many landslides combine two or more of these movements at the same time or during the lifetime of a landslide [28]. Traditionally, landslide inventory maps can be retrieved from historical sources, archives documents, and newspapers, which are important but not detailed enough, making quantitative risk assessment challenging [29]. Historical landslide events offer a good opportunity to evaluate landslide detection techniques to develop landslide inventory, which can also be used in developing or validating landslide susceptibility and hazard mappings models [30].

An inventory map identifies landslides in a study area to establish the spatial correlation between landslides and environmental factors [31]. A landslide inventory map is required to quantify landslide occurrence effectively. Landslide inventories generally include the size of landslides, its locations, and volume (preferred) [26]. Inventory maps, which provide information on the location of landslide distributions, help implement necessary mitigation measures [25]. The first step in evaluating landslide hazards is a comprehensive landslide inventory map [32].

Although landslide inventory mapping is a crucial requirement for a thorough hazard and risk analysis, the usefulness of these maps for land management and planning is rather limited due to their inhomogeneous spatial distribution and the use of different mapping and classification criteria and methods [33]. However, large-scale landslide inventory maps, developed using remotely sensed data, can overcome these problems and limitations, thus making susceptibility hazard and risk assessment more efficacious [34].

Since it is time consuming and expensive to develop landslide inventories using in situ measurements, remote sensing data, and tools can be an effective way to develop landslide inventory maps. High-resolution satellite image and advanced remote sensing and spatial analysis techniques allow developing more reliable landslide inventory maps [29, 35]. Methods involved in generating landslide inventories include visual interpretation of multi-temporal aerial photographs and remotely sensed images and geomorphological field mapping, expert knowledge on the geological setting in combination [36, 37].

For example, Harp et al. [38] suggested mapping criteria for landslide inventory as follows:

a. GIS tool can be used to develop a spatial map for statistical analysis in relation to other spatial variables; and

b. the entire population of landslides triggered by the external forces or natural and anthropogenic causes must be mapped and plotted so that a complete landslide distribution can be obtained.

However, detailed inventories are rarely available because evidence could have been lost due to various degrees of modification in the past [26]. On the other hand, an inventory map/data can be developed using high-resolution historical remote sensing data, especially if morphological indicators for past landslide activities are present [34].

According to Guzzetti et al. [33] and Malamud et al. [26], landslide inventory maps are categorized into archive inventory (based on records in archives, newspapers, and so on). Similarly, in Guzzetti et al. [33] landslide inventory maps are categorized as geomorphological inventory which can further be classified as follows:

- historical inventory (showing the cumulative effect of landslide over a long period without further temporal differentiation),

- event-based inventory (landslides caused by a single triggering event, such as a strong earthquake),

- seasonal inventory (landslides triggered within one active season) and multi-temporal inventory (continuous monitoring of landslide activity over longer periods independent of particular triggering events), and

- the multi-temporal inventory is the most labor-intensive inventory type and the only one with the potential for spatio-temporal completeness, and it generally requires the use of remote sensing.

A historical inventory is the most popular approach developing landslide inventories using past landslide events. High spatial resolution and long-term remotely sensed data help rapid mapping and monitoring, especially during an emergency [1]. Based on the available in situ information and remotely sensed data, various landslide inventories can be produced and detailed at different scales covering the entire area affected by landslides, including, wherever possible, all sizes of landslides, and mapping landslides as polygons to depict their exact shapes [38].

In addition, landslide inventory maps are prepared for multiple scopes, including:

- documenting the extent of landslide phenomena in areas ranging from small to large watersheds and from regions to states or nations,

- taking a preliminary step toward landslide susceptibility, hazard, and risk assessment,

- investigating distribution, types, and patterns of landslides in relation to morphological and geological characteristics, and distribution of slope failure processes,

- studying the evolution of landscapes dominated by mass-wasting processes,

- investigating the recent and historical relationships between mass movement processes, settlements, and high cultural value areas [39], and

- extracting thresholds of rainfall-induced landslides [40].

Landslide inventory and detection are technically very close and often used interchangeably. An inventory must be carried out through direct visual inspections (or field surveys) and/or in situ measurements (when possible), which, altogether, identify "detection." In other words, there is no good and effective inventory without detection.

A complete and accurate landslide inventory is crucial for landslide predictions; therefore, the accuracy of landslide inventory can be maintained by analyzing high-resolution satellite images [41]. Identification of landslide's boundaries, terrestrial and topography verification, and third-party review are the procedures in interpreting the accuracy of landslide inventory [41].

The systematic information on the type, abundance, and distribution of landslide is still lacking, which helps document essential details on landslide types, patterns, recurrence, and statistics of slope failures. These pieces of information are helpful to identify landslide susceptible zones and determine potential hazards and landslide risk assessment [33]. Multi-temporal inventories are needed, especially in regions of frequently occurring landslide. These regions require high spatio-temporal resolution data and efficient methods for landslide mapping and analysis [42]. An inventory map which contains landslide type, state of activity, depth, volume, date, and place of occurrence, can be used for the calculation of predisposing factors for performance and reliability analysis [30, 37]. In addition to analyze and understand the causes of past landslides, landslide detection is equally useful for monitoring and predicting future hazards [43].

Landslide inventory/detection maps can be developed using the consolidated procedure of photo-interpretation of different sets of stereoscopic aerial photos, that can be integrated with an extensive field control of each recognized landslide. The field control process includes acquiring additional information about the main geomorphic elements and topographic signatures related to mass movement processes and their interpretation in terms of pattern, distribution, state of activity, and evolution of slope processes. In particular, the geomorphological field survey focuses on:

1. the validation of the information acquired by aerial photo-interpretation,

2. recognition of landslide types and state of activity,

3. analysis of deposits involved in slope failures, and

4. evaluation of damages to infrastructures.

Landslides of different sizes and types also offer to detect landslides, which can be used to develop a landslide inventory map [30]. A landslide can be detected or identified using visual interpretation techniques combined with field investigations as a ground control to develop the most reliable form of inventory maps for scientific studies [44–47].

Numerous researchers have conducted studies to develop landslide inventories using remotely sensed data (e.g., [35, 37, 48–50]) (**Table 1**). For example, Moosavi et al. [37] produced a landslide inventory map using GeoEye remotely sensed data and found that thematic mapping using high spatial resolution satellite imagery necessitates a new methodology. Also, Sun et al. [35] used Geofen-1 remotely sensed data to develop a landslide inventory map in a complex region with numerous gullies, which was otherwise challenging and impractical via field investigations.

Satellites	Spatial resolutions (m)	Launch date	Applications	References for examples
Geofen-1	2	2013	Inventory	[35]
GeoEye-1	1.65	2008	Inventory	[37]
IRS-1C Resources at 2-LISS III RapidEye	188 24 5	1995 2011 2008	Inventory	[48]
TERRA/ASTER	15, 30	1999	Inventory	[49, 50]
SPOT-5	2.5 Pan 10 Mult	2002	Detection Mapping	[39, 51]
Quickbird-2	0.6	2001	Detection	[52, 53]
LiDAR	0.5	NA	Detection	[33, 45, 54]
IKONOS	0.82 Pan, 3.2 Mult	1999	Mapping	[55]
Landsat-8	15, 30	2013	Mapping	[56, 57]
SRTM	30, 90	2000	Susceptibility Detection	[16, 57]
Cartosat-2	1.0	2017	Susceptibility	[25]
Cartosat-1	2.5	2005	Susceptibility	[30]
AMSR-E	25,000	2002	Susceptibility	[16]
TRMM	25,000	1997	Susceptibility	[58]
GPM-Integrated	Approx. 11,000	NA	Susceptibility	[59, 60]
WorldView-1/ Quickbird-2/ GeoEye-1	1.85/2.4/1.65	2009/2001/2008	Susceptibility Inventories	[61]
Landsat MSS Landsat TM	80 30	1984	Interpretation	[4, 62]
Landsat	30	1992	Planning	[63]
Landsat 7	15, 30	1999	Characterizing Identification	[64, 65]
ALOS ALOS/PALSAR	10	2007	Mapping Deformation	[1, 52]
Landsat TM/SPOT-5	15, 30/2.5, 10	1999/2002	Hazard	[66]
IRS/Landsat-7	6, 23.6, 188/30	1997/1999	Hazard	[67]
Radarsat-1 ERS-2	8 10	1995 1995	Characterizing Monitoring	[68]
ERS SAR	10	1995	Monitoring	[69]
ERS-1/2	10	1991/1995	Monitoring	[70]
COSMO-SkyMed-1, 3	1, 5, 15, 30, 100	2007/2008	Detection	[71]
TerraSAR-X	1	2007	Monitoring	[72]

This list is an effort to compile popularly used satellite data to study landslides; it is not meant to be a comprehensive list as there are many more studies that used other satellite products.

Table 1.
A series of remotely sensed data used for landslide study.

3. Landslide susceptibility mapping

Landslide prediction is vital to prevent possible damages and save human lives. Landslide susceptibility map is important in predicting landslides because it helps to identify potential landslide areas and any area susceptible to landslides [30, 73, 74]. The local topography and hydrological conditions play a significant role in landslide susceptibility [39]. Although a proper landslide inventory may provide both spatial and temporal information on previous landslides over an area, landslide susceptibility map gives information about potential future landslides over an area [60]. However, detailed information on the historical records of previous landslides, rainfall, or earthquake is vital in determining triggering thresholds [74].

Landslide susceptibility can be quantified from stable to highly susceptible. Many researchers categorize slopes into four landslide susceptibility classes; highly susceptible, moderately susceptible, slightly susceptible, and stable (e.g., [12, 16, 58, 75–78]). Some researchers used slightly different susceptible classes to develop landslide susceptibility maps such as unstable, quasi-stable, moderately stable, and stable (e.g., [12]). Some studies also used susceptibility indices: very high, high, moderate, low, and very low (e.g., [59–61, 79]).

For disaster prevention, a landslide susceptibility map can be used in land-use planning and decision making [39]. A detailed susceptibility map for land-use mapping helps local authorities manage these landscapes for urban or industrial planning and development [80]. However, developing an effective landslide susceptibility map is always a challenge because it requires multiple spatial information of soil, geology, vegetation, and hydrology. For example, Stanley and Kirschbaum [60] identified four major issues that are important to be addressed for developing landslide susceptibility maps:

i. lack of detailed inventories,

ii. minimum available input data,

iii. regional differences in the importance of causative factors, and

iv. the dearth of expertise on landscape processes across large regions.

Numerous methods exist in literature developing landslide susceptibility map using in situ measurements, models, remotely sensed data either stand-alone or in combination. For example, landslide inventories and causative factors, along with the statistical approach, are used in developing landslide susceptibility model for predicting potential landslides [39]. In addition, many studies used statistical approaches, physically based models, and deterministic approaches in developing landslide susceptibility maps (e.g., [53, 62, 75, 81]).

Besides the different existing methods for landslide susceptibility analysis and mapping, it is also essential to have advanced tools and detailed spatial information to develop an effective landslide susceptibility map. Digital tools such as GIS and global positioning system (GPS) are mostly used to analyze spatial data and developing landslide susceptibility and hazard maps [29, 81]. Moreover, remotely sensed data and technologies are widely used for effective landslide susceptibility mapping, hazard assessment, and risk assessment, which further helps in awareness, mitigation, and management of potential landslide threats [29].

Many researchers have used remotely sensed data to develop landslide susceptibility maps from local to global scale. For example, Ray and Jacobs [59] and

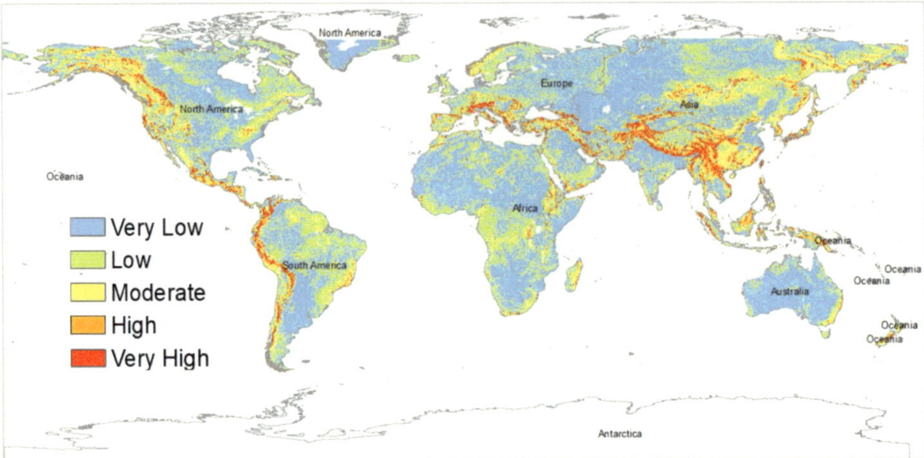

Figure 1.
Global susceptibility map (slightly modified from Stanley and Kirschbaum [60]).

Kirschbaum et al. [60] used remotely sensed precipitation data [TRMM and global precipitation measurement (GPM)] combined with slope, geology, road networks, fault zones, and forest loss to develop landslide susceptibility map at the global scale (**Table 1**, **Figure 1**). Ray et al. [16] used remotely sensed soil moisture (AMSR-E) combined with slope, soil, and vegetation characteristics to develop dynamic landslide susceptibility maps at a regional scale.

4. Landslide hazard analysis

Landslide is a major natural hazard that leads to a significant loss to human lives and properties. Landslide hazard requires systematic and objective assessment of the multi-landslide hazard, which includes different characteristics and casual factors of hazard along with their spatial, temporal, and size probabilities [48, 82]. Effective planning and management reduce social and economic losses caused by landslides [30, 83, 84]. A landslide susceptibility map combined with temporal information can be converted into a landslide hazard map for estimating potential losses due to landslides [30]. Landslide risk can be estimated using landslide hazard maps [85]. A useful hazard map should include local geology, geomorphology, lithology, hydrology, vegetation, and climatic factors. These factors affect landslide events, needed for proper hazard analysis [86].

A vital part of hazard assessment is the quantitative estimate of the pre-failure and failure stages of the susceptibility of the slope [87]. Landslide hazard assessment determines slope failure probability [88]. Over the last 30 years, numerous studies have been conducted on landslide hazard zonation as a result of the demand for slope instability hazard for planning purposes [89]. Despite that susceptible slopes triggers or reactivate slope failures, hazard analysis must consider the speed of landslide movement along the slope [88]. Huabin et al. [90] suggested two important aspects of landslide hazard zonation, which are assessing the susceptibility of the terrain for slope failure, and determining the probability of a specific triggering event occurring.

Nadim and Kjekstad [91] used a landslide hazard index (Hlandslide) to classify landslide hazard levels from negligible to very high (**Table 2**). The Hlandslide was obtained by multiplying a series of factors such as slope factor (Sr), lithology

Value of $H_{landslide}$	Class	Classification of landslide hazard potential	Approximate annual potential frequency in 1 km^2 grid
<14	1	Negligible	Virtually zero
15–20	2	Very low	Negligible
51–100	3	Low	Very small
101–168	4	Low to moderate	Small
169–256	5	Moderate	0.0025–0.01%
257–360	6	Medium	0.0063–0.025%
360–512	7	Medium to high	0.0125–0.05%
513–720	8	High	0.025–0.1%
>720	9	Very high	0.05–0.2%

Table 2.
Landslide hazard index ($H_{landslide}$), adapted from Nadim and Kjekstad [91].

factor (Sl), soil moisture factor (Sh), precipitation trigger factor (Tp), and seismic trigger factor (Ts); the following equation was used to develop Hlandslide.

$$H_{landslide} = \left(S_r * S_l * S_h\right) * \left(T_s + T_p\right) \quad (1)$$

where S_r = slope factor, S_l = lithology factor, S_h = soil moisture factor, T_s = seismic trigger factor, and T_p = precipitation trigger factor.

Remotely sensed data provide several essential factors used in the equation to develop landslide hazard zones on the range of scales. It is often difficult to obtain the multi-spatio-temporal information on landslide occurrences needed for landslide hazard analysis [48]. Chau et al. [86] suggested that landslide analysis should include landslide-dynamics-based numerical simulations to prevent subjectivity and bias; incorporation with GIS should result in an adequate hazard map to work for better planning.

Susceptible areas can be assessed and predicted, thereby reducing damage caused by landslides through proper preparation and mitigation because landslide prevention is a severe challenge [4]. Generally, landslide hazard analysis is conducted using aerial photographs, and/or remotely sensed images; therefore, it might contain a large degree of uncertainty (**Table 3**) [92]. As listed in **Table 2**, the degree of uncertainty is related to many factors, such as topography, soil, vegetation, and hydrology. On the other hand, the level of uncertainty is strongly related to the degree of susceptibility of a map [92]. Also, the probability of landslide hazard depends on both the intrinsic and extrinsic variables. Intrinsic variables include geological conditions and slope structures, whereas extrinsic variables include rainfall and human activities [90]. Chau et al. [86] explained that a reliable landslide hazard map should include historical landslide events, geomorphological analysis, and mechanical analysis of slides, falls, and flows of earth mass. Since all three aspects of hazard analysis involve handling and interpreting a large amount of data, spatial analysis tools such as GIS is essential for such analysis.

Spatial information from previous landslide events is needed for landslide analysis and evaluating the probability of future landslide occurrence [48]. Therefore, high-resolution spatial information (satellite data) for factors associate in landslide hazards is essential for effective hazard analysis.

Factors	Uncertainty	Factors	Uncertainty
Slope angle	Low	Rainfall distribution	Intermediate
Slope direction	Low	Morphological setting	Low
Slope convexity	Low	Detailed geomorphological situation	Intermediate/high
General lithological zonation	Low	Present mass movement distribution	Intermediate
General lithological composition	High	Present mass movement typology	Intermediate
General tectonic framework	Low	Present mass movement activity	Intermediate/high
Detailed rock structure	High	Past mass movement distribution	High
Soil type distribution	Low/intermediate	Land use	Low
Soil characteristics	Intermediate/high	Past climatological condition	High
Soil thickness	High	Earthquake acceleration	High
Groundwater conditions	High		

Table 3.
Main factors in landslide hazard zonation and their estimated degree of uncertainty, adapted from Mantovani et al. [92].

Recently, GIS tools and remotely sensed data have proven a vital approach for comparing and analyzing landslide, whereby a probabilistic landslide hazard analysis for the affected region is produced [38, 90]. Multiple layers of information are incorporated into the GIS system for a more accurate and reliable landslide hazard and risk analysis [86]. Geotechnical and safety factor-based models are also recommended for an effective landslide hazard analysis [4]. Various scenarios with different volumes or sliding surfaces should be integrated for hazard analysis of potential unknown landslides [2].

Golovko et al. [48] used multiple satellite data (e.g., Landsat, Spot, Aster, IRS-1C, LISS III, and RapidEye) and automated detection techniques to develop landslide susceptibility map and landslide hazard index. They summarized that their presented approach was based on the extensive use of remote sensing data and geospatial tools (e.g., GIS) to characterize landslide susceptibility and hazard. Ray et al. [16] used satellite soil moisture and hydrologic model in combination to develop landslide susceptibility maps at active Cleveland Corral landslide area in California, U.S. Ray et al. [76] used an integrated approach to combine satellite soil moisture and a hydrologic model to develop susceptibility maps at Dhading, Nepal.

5. Landslide monitoring

Landslide monitoring includes all of the activities discussed earlier, such as developing landslide inventory/detection, landslide susceptibility maps, and conducting a landslide hazard analysis. The easiest way to provide geological information to decision-makers and the public is through maps or visualization, which show locations of the landslide events, or where it might occur, thereby providing information on landslide hazard zones [37]. The most effective way to minimize and reduce the impact of landslide hazards and improve risk management is through landslide monitoring and planning [30, 35].

Landslides occur due to the combined effect, such as intense rainfall, topography, and antecedent soil moisture [93]. However, landslides can also be triggered due to external driving forces, such as earthquake, volcano, and excessive surcharge load on the slopes. Landslide hazard causes enormous infrastructural damages and human casualties in mountainous regions [20], and environmental degradation [94]. The combined effect of surface and sub-surface saturation is critical because landslide trigger is not due to only surface layer saturation [92]. Landslide monitoring technique depends on the type and size of the landslide, and the risks involved; it also differs between countries because of their available technology and expertise in landslide monitoring, past experiences with a landslide, and other factors [94]. For example, monitoring the surface displacements of a slope provides essential data for landslide dynamics [94].

Landslide hazard mitigation measures include hazard mapping and assessment, real-time monitoring systems of active and complex landslides, and emergency planning [94]. Landslide monitoring includes a comparison of the areal extent, speed of debris movement, rate of slope movement, surface topography, etc., concerning landslide conditions from different landslide occurrences to assess the activity of a landslide [92]. Timely and high-quality information received from spatial observations is crucial for managers of natural and human-made disasters, particularly in response to emergencies [1]. Landslide monitoring has improved over the years, with better monitoring equipment, automatic measurements done by machines, and less expensive tools [95].

Remote sensing data and techniques can be used for in-depth hazard mapping and monitoring because of their extensive coverage and frequency of observations, especially in high mountainous regions [20, 41, 92, 96]. Stereoscopic air-photo interpretation as far back as the late 1990s has been the most consistent remote sensing tool for landslide monitoring [20]. Combining aerial photography and infrared imagery gives a better result of terrain conditions rather than from either system separately [94].

Coupled with pre-existing landslide inventory maps and synthetic aperture radar (SAR) imagery and interferometric synthetic aperture radar (InSAR) through the integration of auxiliary data (e.g., detailed geological information) can be an effective method to update landslide inventory [1, 20, 97]. Ground-based optical systems (video cameras) limits monitoring movement of active landslides, in case of fog, rain, and darkness [94]. InSAR is useful to monitor prolonged slope movements [20]. Also, InSAR has been widely used in research because of its broad coverage, high spatial resolution, and ability to operate under all weather conditions [52].

Landslide monitoring includes detailed information on topography, geology, groundwater levels, material properties, possible mass movements [93]. Several types of instruments, such as extensometers, inclinometers, piezometers, strainmeters, pressure cells, geophones, tiltmeters, and crack meters, have been used to monitor slope movements and deformation [92, 93]. Recently, the landslide monitoring system has been improved with the development of less expensive computerized equipment [95]. According to Savvaidis [94], and Macek et al. [95], landslide monitoring systems and techniques include:

1. Remote sensing techniques with space-derived information. They are of importance for seismic hazards, landslide hazards, and management of earthquake disasters. Also, remote sensing tools provide surface soil moisture data up to 1–5 cm deep, which has been used to develop landslide susceptibility maps [93].

2. Global positioning system (GPS) technique. It is a flexible and easy operation that uses a series of satellites to determine accuracy in the order of centimeters [92]. A handled GPS provides accurate differential positioning over several kilometers [94].

3. Photogrammetric technique, combined with digital imaging sensors data. It allows early identification of landslide hazard [96]. A photogrammetric technique, which includes interpretation of aerial photography, is a useful technique to identify and map landslides for an extended period [20, 92, 94, 98]. It is also a valuable technique for identifying and describing the 3D overview of the terrain in determining surface information [96].

4. Ground-based conventional surveying techniques measure for absolute displacement computations with the use of different instruments, usually employed in an episodic `monitoring program [94].

5. Geotechnical methods make use of sensors permanently working on or in the region under consideration, and can also use a telemetric system for real-time transmission of data to a control center [94].

6. Types and role of remote sensing techniques

Several remote sensing data have been used to study landslide processes, including space-borne synthetic aperture radar (SAR) and optical remote sensing, airborne light detection and ranging (LiDAR), ground-based SAR and terrestrial LiDAR, incorporating in situ measurements and observations of environmental factors (**Table 1**). In particular:

- SAR data have been widely used in landslide research because of their broad coverage and high spatial resolution and the ability to operate under all weather conditions. Satellite SAR data used include archived ERS and Envisat ASAR [46], ALOS/PALSAR [52, 99], COSMO-SkyMed constellation [71, 100], TerraSAR-X [72, 101], TerraSAR-X/TanDEM-X [100] and Sentinel-1 [102] and Envisat 2010+ data (22 October 2010–8 April 2012).

- Optical remote sensing images were mainly applied to generate landslide inventory, considering long time-series of Landsat TM/ETM, SPOT 1–5, ASTER, IRS-1C LISS III, and RapidEye between 1986 and 2016 [48]. The ZY-3 [103], and GF-1 [35] high spatial resolution satellite images were used to investigate the landslide cinematics with an image correlation algorithm to SPOT-5 images [104].

- Multi-temporal LiDAR images and ortho-photos can be compared to quantify landscape changes caused by an active landslide [105]. The ground-based terrestrial laser scanner (TLS) LiDAR can produce highly detailed three dimensional (3D) images within minutes, allowing the study of 3D surface changes of landslides [106].

Among the most useful applications derived from the analysis of remote sensing images is the development of digital terrain models (DEMs), such as those generated from Indian remote sensing satellite (IRS) P5 images [107] and TerraSAR-X/TanDEM-X images by InSAR [108]. DEM can then be used to assess erosion, landslide, and topographic multi-temporal differences [107].

Lu et al. [45] used Quickbird remotely sensed data for landslide detection and mapping. They summarized that traditional mapping techniques for landslide detection and mapping, which rely on manual interpretation of aerial photographs and intensive field surveys, are time consuming and not efficient for generating such event-based maps. Guzzetti et al. [33] used aerial photographs, high-resolution DEM (LiDAR), and satellite images (e.g., Landsat-7, IRS, IKONOS-2, Quickbird-2, WorldView-2, and GeoEye-1/2) to develop multi-temporal landslide inventory maps. Holbling et al. [51] used SPOT-5 remotely sensed data for landslide change detection, whereas [52] used ALOS/PALSAR imageries and InSAR techniques for landslide detection. On the other hand, Desrues et al. [54] used LiDAR DEM and satellite images for landslide detection and mapping.

7. Global application

There are several studies conducted using remote sensing data and technologies around the globe. This chapter intends to summarize some of the remote-sensing based research undertaken to address landslide issues from selected countries, which are, mainly, more vulnerable to landslides. The studies have shown that several remote sensing techniques can assist in producing landslide inventory and risk assessment maps by providing information on the morphological features of landslides.

These approaches are very useful mainly in very large geographical areas where landslides are the most common yet highly devastating disasters, such as in Nepal, U.S., Philippines, and in many other countries. In the U.S., landslides caused 25 to 50 deaths each year, whereas extreme rainfall is the most common cause of landslides in the Philippines [109]. Ray and Jacobs [93] studied landslides in California, U.S., Leyte, Philippines and, Dhading, Nepal. They established the relationship between landslides, satellite soil moisture (Advanced Microwave Scanning Radiometer (AMSR-E)), and satellite precipitation (Tropical Rainfall Measuring Mission (TRMM)). In Nepal, Amatya et al. [61] used high resolution optical data for landslide mapping and susceptibility analysis along the Karnali Highway in Nepal.

Light detection and ranging (LiDAR) data open unprecedented possibilities for landslide mapping, with potential opportunities for hazard and risk zonation and landscape evolution modeling [33]. Gorsevki et al. [110] used LiDAR data to detect landslides in the Cuyahoga Valley National Park, Ohio, U.S, to generate a susceptibility map using the artificial neural network (ANN). Martha et al. [30] used a semiautomatic approach to develop landslide inventories from post-event satellite images, which they used for landslide susceptibility, hazard, and risk in the High Himalayan terrain in India. Following the 2004 Typhoon Aere, the object-based image analysis approach (OBIA) was adopted to develop landslide inventory in Xiulan, Taiwan [111].

Landslide mapping in the Cameron Highlands area in Malaysia is complicated due to dense vegetation and weather conditions. However, researchers used airborne synthetic aperture radar (AIRSAR) and WorldView-1 satellite data to develop a landslide inventory map for the Cameron Highlands [112, 113]. Also, Bui et al. [34] used synthetic aperture radar (SAR) data for landslide detection and susceptibility mapping in Cameron Highlands, Malaysia. In China, an area near the Three Gorges Reservoir (TGR) along the Yangtze River, which is one of the most landslide-prone regions in the world, was studied using ZY-3 high resolution satellite images to develop landslide inventory map [103].

In Italy, rainfall-induced landslides are serious threats to the population [114]. Since landslides are a national threat and distributed throughout the nation, it is not possible to analyze and monitor landslides using in situ measurements alone. Several researchers have studied landslides using remotely sensed data and technologies (e.g., [102, 115, 116]). Several researchers have also used cartographic thematic and optical data with Persistent Scatterer Interferometry (PSI) to identify slow to very slow moving landslides in a different region of Italy [117–122]. Boni et al. [115] developed a new methodology to update the landslide state of activity (LAMBDA) using multi-temporal A-DInSAR data at north-western Italy.

Bozzano et al. [46] used field surveys and remote sensing techniques to investigate more than 90 landslides affecting a small river basin in Central Italy. Bardi et al. [116] used GB-InSAR data to monitor the rapid movement of earth flows over the Capriglio landslide in the northern Apennines, Italy. InSAR data were used to conduct a multi-temporal assessment of landslide activity in the basin of Abruzzi, Italy [44]. Fayne et al. [57] used Landsat-8 imagery in identifying potential locations and timing of newly triggered landslides in Italy.

Satellite-born landcover panchromatic images and shuttle radar topography (SRTM) elevation data were used to monitor landslide clusters over a vegetated landscape in Itaoca, Brazil [123]. Landslide is one of the natural hazards that occur each year in Indonesia. Hayati et al. [124] used interferometric synthetic aperture radar (InSAR) data to monitor the slow-moving landslide in Ciloto, Indonesia. The sensors provided useful data in obtaining continuous and area-based information for landslide affected areas. Bravo-Carvajal et al. [125] used SPOT satellite images to develop landslide inventory and susceptibility maps at the village of Nueva Colombia in the state of Chiapas. Also, in northern Pakistan, the SPOT-5 satellite was used to develop a landslide inventory [126]. Shroder and Weihs [127] used ASTER elevation data and Landsat 7 images for landslide mapping at north-eastern Badakhshan, Afghanistan. Rainfall-induced landslides are frequent in the mountainous region of Bangladesh. For example, Das and Raja [128] used ASTER elevation data and Landsat 8 images to develop a landslide susceptibility map at Chittagong city, Bangladesh. Singhroy et al. [64] used Landsat TM and SAR images in combination to interpret the retrogressive slope failures on the Shale banks of the Saskatchewan river in Canada. Singhroy et al. [64] used SAR images to identify flow slides on sensitive marine clays in the Ottawa Valley of Canada.

8. Summary and conclusions

Natural disasters such as hurricanes, earthquakes, tsunami, and landslides have been on the rise, causing damage to property and human lives, especially in mountainous regions. Major causes of landslides are conditioning factors, such as lithology, relief, geological structure, geomechanical properties, weathering, and triggering factors such as precipitation, seismicity, temperature change, and static and dynamic loads. Conventional methods for landslide studies mainly rely on the visual interpretation of aerial photographs and field investigation in combination. However, these methods are time consuming and cost-ineffective. On the other hand, remotely sensed data at high spatial and temporal resolutions and advanced techniques could be used for landslide studies at a range of scales, which can reduce the time and resources required for the studies.

This chapter has reviewed the use of remotely sensed data and advanced technologies used for landslide inventory/detection, developing landslide susceptibility maps, conducting a landslide hazard analysis, and monitoring landslide events and slope movement at a range of scales. More robust technologies and high-resolution data are essential to reduce the impact of landslides threats on human lives, properties, and environments. In addition, we also need to develop advanced technologies, which can improve landslide assessment, prediction, and mitigations. According to Singhroy [129], our primary challenge is to have an advanced technology and/or high-resolution data to recognize and interpret detailed geomorphic characteristics of large and small landslides and determine whether or not failure is likely to occur. Although the use of high spatial resolution radar and LiDAR data are very helpful in conducting landslide studies, satellite products at high spatial and temporal resolutions are still limited. In the future, if real-time remotely sensed products at high spatial and temporal resolutions are available, especially in remote and hardly accessible terrain, it would be helpful to study landslide dynamics at a range of scales.

Acknowledgements

This work was supported by the Evans-Allen project of the United States Department of Agriculture (USDA), National Institute of Food and Agriculture.

Conflict of interest

The authors declare no conflict of interest.

Author details

Ram L. Ray[1*], Maurizio Lazzari[2] and Tolulope Olutimehin[1]

1 College of Agriculture and Human Sciences, Prairie View, Texas, USA

2 CNR-ISPC, C/da S. Loja, Potenza, Tito Scalo, Italy

*Address all correspondence to: raray@pvamu.edu

References

[1] Casagli N, Cigna F, Bianchini S, Hölbling D, et al. Landslide mapping and monitoring by using radar and optical remote sensing: Examples from the EC-FP7 project SAFER. Remote Sensing Applications: Society and Environment. 2016;**4**:92-108. DOI: 10.1016/j.rsase.2016.07.001

[2] Lari S, Frattini P, Crosta GB. A probabilistic approach for landslide hazard analysis. Engineering Geology. 2014;**182**:3-14. DOI: 10.1016/j.enggeo.2014.07.015

[3] Tofani V, Segoni S, Agostini A, Catani F, Casagli N. Technical note: Use of remote sensing for landslide studies in Europe. Natural Hazards and Earth System Sciences. 2013;**13**:299-309. DOI: 10.5194/nhess-13-299-2013

[4] Pradhan B. Remote sensing and GIS-based landslide hazard analysis and cross-validation using multivariate logistic regression model on three test areas in Malaysia. Advances in Space Research. 2010;**45**:1244-1256. DOI: 10.1016/j.asr.2010.01.006

[5] McKean D, Calkins H, Peuquet D. Remote sensing and landslide hazard assessment. Photogrammetric Engineering and Remote Sensing. 1991;**57**(9):1185-1193

[6] Lazzari M, Geraldi E, Lapenna V, Loperte A. Natural hazards vs human impact: An integrated methodological approach in geomorphological risk assessing on Tursi historical site, southern Italy. Landslides. 2006;**3**(4):275-287

[7] Lazzari M, Piccarreta M. Landslide disasters triggered by extreme rainfall events: The case of Montescaglioso (Basilicata, southern Italy). Geosciences. 2018;**8**(10):377

[8] Guzzetti F, Reichenbach P, Cardinali M, Galli M, Ardizzone F. Probabilistic landslide hazard assessment at the basin scale. Geomorphology. 2005;**72**:272-299. DOI: 10.1016/j.geomorph.2005.06.002

[9] USGS. Landslide Types and Processes. 2004. Available from: https://pubs.usgs.gov/fs/2004/3072/fs-2004-3072.html [Accessed: 27 April 2020]

[10] Pathak D. Remote sensing and GIS application in landslide risk assessment and management. Nepalese Journal of Geoinformatics. 2016;**15**:52-57

[11] Gupta M, Ghose MK, Sharma LP. Application of remote sensing and GIS for landslides hazard and assessment of their probabilistic occurrence A case study of NH31A between Rangpo and Singtam. Journal of Geomatics. 2009;**3**(1):13-17

[12] Ray RL, De Smedt F. Slope stability analysis on a regional scale using GIS: A case study from Dhading, Nepal. Environmental Geology. 2009;**57**:1603-1611. DOI: 10.1007/s00254-008-1435-5

[13] Lombardo L, Martin Mai P. Presenting logistic regression-based landslide susceptibility results. Engineering Geology. 2018;**244**:14-24. DOI: 10.1016/j.enggeo.2018.07.019

[14] Flentje PN, Miner A, Whitt G, Fell R. Guidelines for landslide susceptibility, hazard and risk zoning for land use planning. Australian Geomechanics Journal. 2007;**42**(1):13-36

[15] Ciampalini A, Raspini F, Lagomarsino D, Catani F, Casagli N. Landslide susceptibility map refinement using PSInSAR data. Remote Sensing of Environment. 2016;**184**:302-315. DOI: 10.1016/j.rse.2016.07.018

[16] Ray RL, Jacobs JM, Cosh MH. Landslide susceptibility mapping using

downscaled AMSR-E soil moisture: A case study from Cleveland corral, California, US. Remote Sensing of Environment. 2010;**114**:2624-2636. DOI: 10.1016/j.rse.2010.05.033

[17] Lazzari M, Gioia D, Anzidei B. Landslide inventory of the Basilicata region (Southern Italy). Journal of Maps. 2018;**14**(2):348-356

[18] Abdallah C, Chowrowicz J, Bouheir R, Dhont D. Comparative use of processed satellite images in remote sensing of mass movements: Lebanon as a case study. International Journal of Remote Sensing. 2007;**28**(19):4409-4427. DOI: 10.1080/01431160701241761

[19] Tsai F, Hwang J-H, Chen L-C, Lin T-H. Post-disaster assessment of landslides in southern Taiwan after 2009 typhoon Morakot using remote sensing and spatial analysis. Natural Hazards and Earth System Sciences. 2010;**10**:2179-2190. DOI: 10.5194/nhess-10-2179-2010

[20] Metternicht G, Hurni L, Gogu R. Remote sensing of landslides: An analysis of the potential contribution to geo-spatial systems for hazard assessment in mountainous environments. Remote Sensing of Environment. 2005;**98**:284-303. DOI: 10.1016/j.rse.2005.08.004

[21] Sarkar S, Kanungo DP. An integrated approach for landslide susceptibility mapping using remote sensing and GIS. Photogrammetric Engineering and Remote Sensing. 2004;**70**(5):617-625. DOI: 10.14358/PERS.70.5.617

[22] Pradhan B, Singh RP, Buchroithner MF. Estimation of stress and its use in evaluation of landslide prone regions using remote sensing data. Advances in Space Research. 2006;**37**:698-709. DOI: 10.1016/j.asr.2005.03.137

[23] Gorsevski PV, Gessler PE, Jankowski P. Integrating a fuzzy k-means classification and a Bayesian approach for spatial prediction of landslide hazard. Journal of Geographical Systems. 2003;**5**:223-251. DOI: 10.1007/s10109-003-0113-0

[24] Scaioni M, Longoni L, Melillo V, Papini M. Remote sensing for landslide investigations: An overview of recent achievements and perspectives. Remote Sensing. 2014;**6**:9600-9652. DOI: 10.3390/rs60x000x

[25] Khadka A, Dhakal S, Budha PGIS. Based landslide susceptibility mapping along the road section from Bandeu to Barahabise, Sindhupal Chowk District of Nepal. IJSR. 2018;**7**:465-471. DOI: 10.21275/ART20192474

[26] Malamud BD, Turcotte DL, Guzzetti F, Reichenbach P. Landslide inventories and their statistical properties. Earth Surface Processes and Landforms. 2004;**29**:687-711. DOI: 10.1002/esp.1064

[27] Naudet V, Lazzari M, Perrone A, Loperte A, Piscitelli S, Lapenna V. Integrated geophysical techniques and geomorphological approach to investigate the snowmelt-triggered landslide of Bosco Piccolo village (Basilicata, southern Italy). Engineering Geology. 2008;**98**(3-4):156-167

[28] Cruden DM, Varnes DJ. Landslide types and processes. Chapter 3 in landslides-investigation and mitigation. In: Turner AK, Schuster RL, editors. Transportation Research Board Special Report 247. USA: National Research Council; 1996. pp. 36-75

[29] van Westen CJ, van Asch TWJ, Soeters R. Landslide hazard and risk zonation—Why is it still so difficult? Bulletin of Engineering Geology and the Environment. 2006;**65**:167-184. DOI: 10.1007/s10064-005-0023-0

[30] Martha TR, van Westen CJ, Kerle N, Jetten V, Kumar KV. Landslide hazard and risk assessment using semi-automatically created landslide inventories. Geomorphology. 2012;**184**:139-150. DOI: 10.1016/j.geomorph.2012.12.001

[31] Pellicani R, Spilotro G. Evaluating the quality of landslide inventory maps: Comparison between archive and surveyed inventories for the Daunia region (Apulia, Southern Italy). Bulletin of Engineering Geology and the Environment. 2015;**74**:357-367. DOI: 10.1007/s10064-014-0639-z

[32] Booth AM, Roering JJ, Perron JT. Automated landslide mapping using spectral analysis and high-resolution topographic data: Puget Sound lowlands, Washington, and Portland Hills, Oregon. Geomorphology. 2009;**109**:132-147. DOI: 10.1016/j.geomorph.2009.02.027

[33] Guzzetti F, Mondini AC, Cardinali M, Fiorucci F, Santangelo M, Chang K-T. Landslide inventory maps: New tools for an old problem. Earth-Science Reviews. 2012;**112**:42-66. DOI: 10.1016/j.earscirev.2012.02.001

[34] Bui Tien D, Shahabi H, Shirzadi A, Chapi K, et al. Landslide detection and susceptibility mapping by AIRSAR data using support vector machine and index of entropy models in Cameron highlands, Malaysia. Remote Sensing. 2018;**10**:1527. DOI: 10.3390/rs10101527

[35] Sun W, Tian Y, Mu X, Zhai J, Gao P, Zhao G. Loess landslide inventory map based on GF-1 satellite imagery. Remote Sensing. 2017;**9**:314. DOI: 10.3390/rs9040314

[36] Golovko D, Roessner S, Behling R, Wetzel H-U, et al. Development of multi-temporal landslide inventory information system for Southern Kyrgyzstan using GIS and satellite remote sensing. PFG. 2015;**2**:157-172

[37] Moosavi V, Talebi A, Shirmohammadi B. Producing a landslide inventory map using pixel-based and object-oriented approaches optimized by Taguchi method. Geomorphology. 2014;**204**:646-656. DOI: 10.1016/j.geomorph.2013.09.012

[38] Harp E, Keefer D, Sato H, Yagi H. Landslide inventories: The essential part of seismic landslide hazard analyses. Engineering Geology. 2011;**122**:9-21. DOI: 10.1016/j.enggeo.2010.06.013

[39] Lazzari M, Piccarreta M, Capolongo D. Landslide triggering and local rainfall thresholds in Bradanic Foredeep, Basilicata region (southern Italy). Landslide science and practice. Volume 2. Early warning, instrumentation and modeling. In: Margottini et al., editors. Proceedings of the Second World Landslide Forum, 3-9 October 2011. Rome, Italy: Springer Series; 2013. pp. 671-678

[40] Lazzari M, Gioia D. UAV images and historical aerial-photos for geomorphological analysis and hillslope evolution of the Uggiano medieval archaeological site (Basilicata, southern Italy). Geomatics, Natural Hazards and Risk. 2017;**8**:104-119

[41] Cheng C-T, Huang C-M, Wei L-W, Lee C-F, Lee C-T. Landslide susceptibility map. In: Sassa K, He B, McSaveney M, Nagai O, editors. International Consortium of Landslides (ICL) Landslide Teaching Tools. Chapter: Part 1. Mapping and Site Prediction. ICL & UNESCO; 2013. pp. 50-55. DOI: 10.1007/s10346-013-0460-y

[42] Behling R, Roessner S, Kaufmann H, Kleinschmit B. Automated spatio-temporal landslide mapping over large areas using RapidEye time series data. Remote Sensing. 2014;**6**:8026-8055. DOI: 10.3390/rs6098026

[43] Li Z, Shi W, Myint S, Lu P, Wang Q. Semi-automated landslide inventory

mapping from bitemporal aerial photographs using change detection and level set method. Remote Sensing of Environment. 2016;**175**:215-230. DOI: 10.1016/j.rse.2016.01.003

[44] Van Den E, Poesen J, Verstraeten G, Vanacker V, et al. The effectiveness of hillshade maps and expert knowledge in mapping old deep-seated landslides. Geomorphology. 2005;**67**:351-363. DOI: 10.1016/j.geomorph.2004.11.001

[45] Lu P, Stumpf A, Kerle N, Casagli N. Object-oriented change detection for landslide rapid mapping. IEEE Geoscience and Remote Sensing Letters. 2011;**8**(4):701-705. DOI: 10.1109/LGRS.2010.2101045

[46] Bozzano F, Mazzanti P, Perissin D, Rocca A. Basin scale assessment of landslides geomorphological setting by advanced InSAR analysis. Remote Sensing. 2017;**9**:267. DOI: 10.3390/rs9030267

[47] Stumpf A, Kerle N. Object-oriented mapping of landslides using random forests. Remote Sensing of Environment. 2011;**115**:2564-2577. DOI: 10.1016/j.rse.2011.05.013

[48] Golovko D, Roessner S, Behling R, Wetzel H-U, Kleinschmit B. Evaluation of remote-sensing-based landslide inventories for hazard assessment in southern Kyrgyzstan. Remote Sensing. 2017;**9**:943. DOI: 10.3390/rs9090943

[49] Gao J, Maro J. Topographic controls on evolution of shallow landslides in pastoral Wairarapa, New Zealand, 1979-2003. Geomorphology. 2010;**114**:373-381. DOI: 10.1016/j.geomorph.2009.08.002

[50] Alkevli T, Ercanoglu M. Assessment of ASTER satellite images in landslide inventory mapping: Yenice-Gökçebey (Western Black Sea region, Turkey). Bulletin of Engineering Geology and the Environment. 2011;**70**:607-617. DOI: 10.1007/s10064-011-0353-z

[51] Hölbling D, Friedl B, Eisank C. An object-based approach for semi-automated landslide change detection and attribution of changes to landslide classes in northern Taiwan. Earth Science Informatics. 2015;**8**:327-335. DOI: 10.1007/s12145-015-0217-3

[52] Kang Y, Zhao C, Zhang Q, Lu Z, Li B. Application of InSAR techniques to an analysis of the Guanling landslide. Remote Sensing. 2017;**9**(10):1046. DOI: 10.3390/rs9101046

[53] Hearn GJ, Hart AB. Landslide susceptibility mapping: A practitioner's view. Bulletin of Engineering Geology and the Environment. 2019;**78**:5811-5826. DOI: 10.1007/s10064-019-01506-1

[54] Desrues M, Lacroix P, Brenguier O. Satellite pre-failure detection and In situ monitoring of the landslide of the tunnel du Chambon, French Alps. Geosciences. 2019;**9**:313. DOI: 10.3390/geosciences9070313

[55] Fiorucci F, Cardinali M, Carlà R, Rossi M, Mondini AC, Santurri F, et al. Seasonal landslide mapping and estimation of landslide mobilization rates using aerial and satellite images. Geomorphology. 2011;**129**:59-70. DOI: 10.1016/j.geomorph.2011.01.013

[56] Wang Q, Wang Y, Niu R, Peng L. Integration of information theory, K-means cluster analysis and the logistic regression model for landslide susceptibility mapping in the three gorges area, China. Remote Sensing. 2017;**9**:938. DOI: 10.3390/rs9090938

[57] Fayne JV, Ahamed A, Roberts-Pierel J, Rumsey AC, Kirschbaum D. Automated satellite-based landslide identification product for Nepal. Earth Interactions. 2019;**23**:1-21. DOI: 10.1175/EI-D-17-0022.1

[58] Ray RL, Jacobs JM. Landslide susceptibility mapping using remotely sensed soil moisture. IEEE Xplore. 2008;**3**:47-50. DOI: 10.1109/IGARSS.2008.4779279

[59] Kirschbaum D, Stanley T. Satellite-based assessment of rainfall-triggered landslide hazard for situational awareness. Earth's Future. 2018;**6**:505-523. DOI: 10.1002/2017EF000715

[60] Stanley T, Kirschbaum DB. A heuristic approach to global landslide susceptibility mapping. Natural Hazards. 2017;**87**:145-164. DOI: 10.1007/s11069-017-2757-y

[61] Amatya P, Kirschbaum D, Stanley T. Use of very high-resolution optical data for landslide mapping and susceptibility analysis along the Karnali highway, Nepal. Remote Sensing. 2019;**11**:2284. DOI: 10.3390/rs11192284

[62] Soeters R, van Westen CJ. Slope instability recognition, analysis and zonation. In: Turner AK, Schuster RL, editors. Landslides: Investigation and Mitigation. Transp. Res. Board. Natl. Acad. Washington, D.C: Press; 1996. pp. 129-177

[63] Bathrellos GD, Gaki-Papanastassiou K, Skilodimou HD, Papanastassiou D, Chousianitis KG. Potential suitability for urban planning and industry development using natural hazard maps and geological–geomorphological parameters. Environment and Earth Science. 2012;**66**(2):537-548. DOI: 10.1007/s12665-011-1263-x

[64] Singhroy V, Mattar KE, Gray AL. Landslide characterisation in Canada using interferometric SAR and combined SAR and TM images. Advances in Space Research. 1998;**21**(3):465-476. DOI: 10.1016/S0273-1177(97)00882-X

[65] Zhou C, Lee C, Li J, Xu Z. On the spatial relationship between landslides and causative factors on Lantau Island, Hong Kong. Geomorphology. 2002;**43**:197-207. DOI: 10.1016/S0169-555X(01)00130-1

[66] Pradhan B, Lee S. Utilization of optical remote sensing data and GIS tools for regional landslide Hazard analysis using an artificial neural network model. Earth Science Frontiers. 2007;**14**(6):143-152. DOI: 10.1016/S1872-5791(08)60008-1

[67] Lee S, Choi J, Min K. Probabilistic landslide hazard mapping using GIS and remote sensing data at Boun, Korea. International Journal of Remote Sensing. 2004;**25**(11):2037-2052. DOI: 10.1080/01431160310001618734

[68] Singhroy V, Molch K. Characterizing and monitoring rockslides from SAR techniques. Advances in Space Research. 2004;**33**:290-295. DOI: 10.1016/S0273-1177(03)00470-8

[69] Berardino P, Costantini M, Franceschetti G, Iodice A, Petranera L, Rizzo V. Use of differential SAR interferometry in monitoring and modelling large slope instability at Maratea (Basilicata, Italy). Engineering Geology. 2003;**68**:31-51. DOI: 10.1016/S0013-7952(02)00197-7

[70] Squarzoni C, Delacourt C, Allemanda P. Nine years of spatial and temporal evolution of the La Valette landslide observed by SAR interferometry. Engineering Geology. 2003;**68**:53-66. DOI: 10.1016/S0013-7952(02)00198-9

[71] Konishi T, Suga Y. Landslide detection using COSMO-SkyMed images: A case study of a landslide event on Kii peninsula, Japan. European Journal of Remote Sensing. 2018;**51**(1):205-221. DOI: 10.1080/22797254.2017.1418185

[72] Zhao F, Mallorqui JJ, Iglesias R, Gili JA, Corominas J. Landslide monitoring using multi-temporal SAR inferometry with advanced persistent scatters identification methods and super high-spatial resolution TerraSAR-X images. Remote Sensing. 2018;**10**:921. DOI: 10.3390/rs10060921

[73] Vakhshoori V, Pourghasemi HR, Zare M, Blaschke T. Landslide susceptibility mapping using GIS-based data mining algorithms. Watermark. 2019;**11**:2292. DOI: 10.3390/w11112292

[74] Fressard M, Thiery Y, Maquaire O. Which data for quantitative landslide susceptibility mapping at operational scale? Case study of the pays d'Auge plateau hillslopes (Normandy, France). Natural Hazards and Earth System Sciences. 2014;**14**:569-588. DOI: 10.5194/nhess-14-569-2014

[75] Ray RL, Jacobs JM, Douglas EM. Modeling regional landslide susceptibility using dynamic soil moisture profiles. Journal of Mountain Science. 2018;**15**(8):1807-1824. DOI: 10.1007/s11629-018-4896-3

[76] Ray RL, Jacobs JM, Ballestero TP. Regional landslide susceptibility: Spatiotemporal variations under dynamic soil moisture conditions. Natural Hazards. 2011;**59**:1317-1337. DOI: 10.1007/s11069-011-9834-4

[77] Acharya G, De Smedt F, Long NT. Assessing landslide hazard in GIS: A case study from Rasuwa, Nepal. Bulletin of Engineering Geology and the Environment. 2006;**65**(1):99-107. DOI: 10.1007/s10064-005-0025-y

[78] Pack RT, Tarboton DG, Goodwin CN. The SINMAP approach to terrain stability mapping. In: Moore D, Hungr O, editors. Proceedings of the 8th IAEG; 21-25 September 1998; Vancouver, British Columbia. Canada: IAEG; 1998. pp. 1157-1165

[79] Pokharel B, Thapa PB. Landslide susceptibility in Rasuwa District of Central Nepal after the 2015 Gorkha earthquake. Journal of Nepal Geological Society. 2019;**59**:79-88. DOI: 103126/jngs.v59i0.24992

[80] Rozos D, Bathrellos GD, Skillodimou HD. Comparison of the implementation of rock engineering system and analytic hierarchy process methods, upon landslide susceptibility mapping, using GIS: A case study from the eastern Achaia County of Peloponnesus, Greece. Environment and Earth Science. 2011;**63**:49-63. DOI: 10.1007/s12665-010-0687-z

[81] Yilmaz I. Landslide susceptibility mapping using frequency ratio, logistic regression, artificial neural networks and their comparison: A case study from Kat landslides (Tokat—Turkey). Computers and Geosciences. 2009;**35**:1125-1138. DOI: 10.1016/j.cageo.2008.08.007

[82] Corominas J, van Westen C, Frattini P, et al. Recommendations for the quantitative analysis of landslide risk. Bulletin of Engineering Geology and the Environment. 2014;**73**:209-263

[83] Dai FC, Lee CF, Ngai YY. Landslide risk assessment and management: An overview. Engineering Geology. 2002;**64**:65-87. DOI: 10.1016/S0013-7952(01)00093-X

[84] Lucieer A, de Jong SM, Turner D. Mapping landslide displacements using structure from motion (SfM) and image correlation of multi-temporal UAV photography. Progress in Physical Geography. 2014;**38**(1):97-116. DOI: 10.1177/0309133313515293

[85] Wu TH, Tang WH, Einstein HE. Landslide hazard and risk assessment. In: Turner AK, Schuster RL, editors. Landslides. Investigation and Mitigation. Transportation Research

Board Special Report. Washington: National Academy Press; 1996. pp. 106-118

[86] Chau KT, Sze YL, Fung MK, Wong WY, et al. Landslide hazard analysis for Hong Kong using landslide inventory and GIS. Computers and Geosciences. 2004;**30**(4):429-443. DOI: 10.1016/j.cageo.2003.08.013

[87] van Asch TWJ, Malet J-P. Flow-type failures in fine-grained soils: An important aspect in landslide hazard analysis. Natural Hazards and Earth System Sciences. 2009;**9**:1703-1711. DOI: 10.5194/nhess-9-1703-2009

[88] Corominas J, Copons R, Vilaplana JM, Altimir J, Amigó J. Integrated landslide susceptibility analysis and hazard assessment in the principality of Andorra. Natural Hazards. 2003;**30**:421-435

[89] van Westen CJ, Terlien MT. An approach towards deterministic landslide hazard analysis in GIS. A case study from Manizales (Colombia). Earth Surface Processes and Landforms. 1996;**21**:853-868

[90] Huabin W, Gangjun L, Weiya X, Gonghui W. GIS-based landslide hazard assessment: An overview. Progress in Physical Geography. 2005;**29**(4):548-567. DOI: 10.1191/0309133305pp462ra

[91] Nadim F, Kjeksta O. Assessment of global high-risk landslide disaster hotspots. In: Sassa K, Canuti P, editors. Landslides disaster risk reduction. Berlin: Springer; 2009. pp. 213-221

[92] Mantovani F, Soeters R, van Westen CJ. Remote sensing techniques for landslide studies and hazard zonation in Europe. Geomorphology. 1996;**15**:213-225. DOI: 10.1016/0169-555X(95)00071-C

[93] Ray RL, Jacobs JM. Relationships among remotely sensed soil moisture, precipitation and landslide events. Natural Hazards. 2007;**43**:211-222. DOI: 10.1007/s11069-006-9095-9

[94] Savvaidis PD. Existing landslide monitoring systems and techniques. In: Proceedings of the Conference from Stars to Earth and Culture; In Honor of the Memory of Professor Alexandros Tsioumis. Thessaloniki, Greece; 2003. pp. 242-258

[95] Maček M, Petkovšek A, Majes B, Mikoš M. Landslide monitoring techniques database. In: Sassa K et al., editors. Landslide Science for a Safer Geoenvironment. Vol. 1. Heidelberg: Springer; 2014. pp. 193-197. DOI: 10.1007/978-3-319-04999-1_24

[96] Kääb A. Photogrammetry for early recognition of high mountain hazards: New techniques and applications. Physics and Chemistry of the Earth, Part B: Hydrology, Oceans and Atmosphere. 2000;**25**(9):765-770. DOI: 10.1016/S1464-1909(00)00099-X

[97] Sun L, Muller J-P. Evaluation of the use of sub-pixel offset tracking techniques to monitor landslides in densely vegetated steeply sloped areas. Remote Sensing. 2016;**8**:659. DOI: 10.3390/rs8080659

[98] van Westen CJ, Getahun FL. Analyzing the evolution of the Tessina landslide using aerial photographs and digital elevation models. Geomorphology. 2003;**1347**:1-13. DOI: 10.1016/S0169-555X(03)00057-6

[99] Qi S, Zou Y, Wu F, Yan C, Fan J, Zang M, et al. A recognition and geological model of a deep-seated ancient landslide at a reservoir under construction. Remote Sensing. 2017;**9**:383

[100] Schlögel R, Thiebes B, Mulas M, Cuozzo G, Notarnicola C, Schneiderbauer S, et al. Multi-temporal X-band radar interferometry using

corner reflectors: Application and validation at the Corvara landslide (Dolomites, Italy). Remote Sensing. 2017;**9**:739

[101] Yang Z, Li Z, Zhu J, Preusse A, Yi H, Hu J, et al. Retrieving 3-D large displacements of mining areas from a single amplitude pair of SAR using offset tracking. Remote Sensing. 2017;**9**:338

[102] Mondini A. Measures of spatial autocorrelation changes in multitemporal SAR images for event landslides detection. Remote Sensing. 2017;**9**:554

[103] Chen T, Trinder JC, Niu R. Object-oriented landslide mapping using ZY-3 satellite imagery, random Forest and mathematical morphology, for the three-gorges reservoir, China. Remote Sensing. 2017;**9**:333

[104] Bivic RL, Allemand P, Quiquerez A, Delacourt C. Potential and limitation of SPOT-5 Ortho-image correlation to investigate the Cinematics of landslides: The example of "Mare à Poule d'Eau" (Réunion, France). Remote Sensing. 2017;**9**:106

[105] Kamps M, Bouten W, Seijmonsbergen A. LiDAR and orthophoto synergy to optimize object-based landscape change: Analysis of an active landslide. Remote Sensing. 2017;**9**:805

[106] Luo L, Ma W, Zhang Z, Zhuang Y, Zhang Y, Yang J, et al. Freeze/thaw-induced deformation monitoring and assessment of the slope in permafrost based on terrestrial laser scanner and GNSS. Remote Sensing. 2017;**9**:198

[107] Ren Z, Zhang Z, Yin J. Erosion associated with seismically-induced landslides in the middle Longmen Shan region, eastern Tibetan plateau, China. Remote Sensing. 2017;**9**:864

[108] Du Y, Xu Q, Zhang L, Feng G, Li Z, Chen R, et al. Recent landslide movement in Tsaoling, Taiwan tracked by TerraSAR-X/TanDEM-X DEM time series. Remote Sensing. 2017;**9**:353

[109] Talubo JP, Jacildo AJ, Espaldon MV, Acosta LA, et al. Vulnerability to rainfall-induced landslide of three communities in Infanta, Quezon, Philippines. International Journal of Sciences: Basic and Applied Research (IJSBAR). 2015;**23**(1):138-166

[110] Gorsevski PV, Brown KM, Panter K, Onasch CM, et al. Landslide detection and susceptibility mapping using LiDAR and an artificial neural network approach: A case study in the Cuyahoga Valley National Park, Ohio. Landslides. 2016;**13**:467-484. DOI: 10.1007/s10346-015-0587-0

[111] Lahousse T, Chang K, Lin Y. Landslide mapping with multi-scale object-based image analysis – A case study in the Baichi watershed, Taiwan. Natural Hazards and Earth System Sciences. 2011;**11**:2715-2726. DOI: 10.5194/nhess-11-2715-2011

[112] Razak K, Santangelo M, Van Westen CJ, Straatsma M, de Jong SM. Generating an optimal DTM from airborne laser scanning data for landslide mapping in a tropical forest environment. Geomorphology. 2013;**190**:112-125. DOI: 10.1016/j.geomorph.2013.02.021

[113] Shahabi H, Hashim M. Landslide susceptibility mapping using GIS-based statistical models and remote sensing data in tropical environment. Scientific Reports. 2015;**5**:9899. DOI: 10.1038/srep09899

[114] Gariano SL, Petrucci O, Guzzetti F. Changes in the occurrence of rainfall-induced landslides in Calabria, southern Italy, in the 20th century. Natural Hazards and Earth System Sciences. 2015;**15**:2313-2330. DOI: 10.5194/nhess-15-2313-2015

[115] Bonì R, Bordoni M, Colombo A, Lanteri L, Meisina C. Landslide state of activity maps by combining multi-temporal A-DInSAR (LAMBDA). Remote Sensing of Environment. 2018;**217**:172-190

[116] Bardi F, Raspini F, Frodella W, Lombardi L, et al. Monitoring the rapid-moving reactivation of earth flows by means of GB-InSAR: The April 2013 Capriglio landslide (Northern Appennines, Italy). Remote Sensing. 2017;**9**:165. DOI: 10.3390/rs9020165

[117] Bianchini S, Cigna F, Righini G, et al. Landslide hotspot mapping by means of persistent scatterer interferometry. Environment and Earth Science. 2012;**67**:1155-1172. DOI: 10.1007/s12665-012-1559-5

[118] Bianchini S, Herrera G, Mateos RM, Notti D, et al. Landslide activity maps generation by means of persistent scatterer interferometry. Remote Sensing. 2013;**5**(12):6198-6222. DOI: 10.3390/rs5126198

[119] Righini G, Pancioli V, Casagli N. Updating landslide inventory maps using persistent Scatterer interferometry (PSI). International Journal of Remote Sensing. 2012;**33**(7):2068-2096. DOI: 10.1080/01431161.2011.605087

[120] Cigna F, Bianchini S, Casagli N. How to assess landslide activity and intensity with persistent scatterer interferometry (PSI): The PSI-based matrix approach. Landslides. 2013;**10**(3):267-283

[121] Lu P, Catani V, Casagli N. Quantitative hazard and risk assessment for slow-moving landslides from persistent scatterer interferometry. Landslides. 2014;**11**(4):685-696

[122] Martire D, Tessitore S, Brancato D, Ciminelli MG, et al. Landslide detection integrated system (LaDIS) based on in-situ and satellite SAR interferometry data. Catena. 2016;**137**:406-421

[123] Batista J, Julien P. Remotely sensed survey of landslide clusters: Case study of Itaoca, Brazil. Journal of South American Earth Sciences. 2019;**92**:145-150. DOI: 10.1016/j.jsames.2019.02.021

[124] Hayati N, Niemeier W, Sadarviana V. Ground deformation in the Ciloto landslides area revealed by multi-temporal InSAR. Geosciences. 2020;**101**:56. DOI: 10.3390/geosciences10050156

[125] Bravo-Carvajal I, Ricardo G, Luis P-B, Irasema A-A. Landslide Susceptibility Mapping Using Remote Sensing and GIS. Nueva Colombia, Chiapas. Mexico: Springer; 2014. DOI: 10.1007/978-3-319-05050-8_63

[126] Khan H, Shafique M, Khan M, Bacha M, et al. Landslide susceptibility assessment using frequency ratio, a case study of northern Pakistan. Egyptian Journal of Remote Sensing and Space Sciences. 2019;**22**:11-24. DOI: 10.1016/j.ejrs.2018.03.004

[127] Shroder JF, Weihs BJ. Geomorphology of the lake Shewa landslide dam, Badakhshan, Afghanistan, using remote sensing data. Geografiska Annaler. Series A, Physical Geography. 2010;**92**(4):469-483. DOI: 10.1111/j.1468-0459.2010.00408.x

[128] Das S, Raja D. Susceptibility analysis of landslide in Chittagong City Corporation area, Bangladesh. International Journal of Environment. 2015;**4**:157-181. DOI: 10.3126/ije.v4i2.12635

[129] Singhroy V. Satellite remote sensing application for landslide detection and monitoring. In: Sassa K, Canuti P, editors. Landslides disaster risk reduction. Berlin: Springer; 2009. pp. 143-158

Chapter 3

Translational Rock-Block Slides in a Tertiary Flyschoid Complexes of Southern Piedmont Region (North-West Italy)

Fabio Luino and Laura Turconi

Abstract

The southern Piedmont Region (north-west Italy) is characterized by a hilly zone called "Langhe" that covers an area of about 2300 km^2 and is bordered by Tanaro River at north and west, by Orba River at east, and by Apennine mountains at south. The Langhe is rolling hills famous for their excellent wine, populated by many small inhabited centers since ancient times. An idea of the Langhe geomorphology can be gained by studying the word "Langa": it may have been derived from either "landa," which means a wild and uninhabited place or from "lingua," which means a strip of land. The morphology of the Langhe hills is characterized by asymmetrical valleys with steep south-east facing slopes and more gentle north-west facing slopes: their profile is defined "saw toothed" by local inhabitants. The asymmetric shape is clearly conditioned by the geology. Severe hydrological events occurred in the last 100 years in Piedmont in particular on May 1926, February and March 1972, February 1974, and November 1994. During these long rainy periods, on the gentler slopes, translational rock-block slides involve tertiary flyschoid complexes represented by rhythmic series of deposits with varied grain size. These landslides often damage or destroy buildings and roads, even if rarely claim human lives.

Keywords: translational rock-block slides, tertiary flyschoid complexes, damage, Piedmont, north-west Italy

1. Introduction

Translational rock-block slides (TRBSs) are mass movements that involve the displacement of material along one or more discrete shearing surfaces. The sliding can extend downward and outward along a broadly planar surface (a translational slide). They are less common than complex landslides, rotational landslides, and shallow landslides. With reference to the TRBSs studied in the world that are not strictly correlated to the Langhe landslides, the casuistry is varied, just as there are many potentially involving rocks. The TRBS phenomenon has been addressed by several authors who have studied them in very different geological contexts due to various trigger causes.

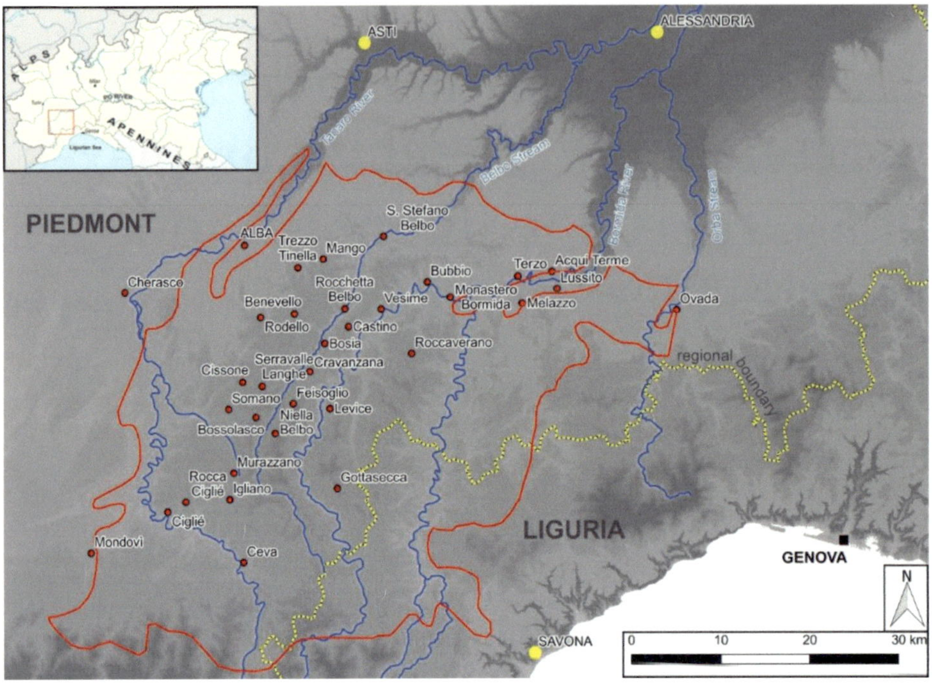

Figure 1.
Langhe hills, a large area in Southeast Piedmont (northern Italy). All place names mentioned in the text and in the captions are marked with colored dots (red limit from Ref. [1], modified).

The singularity of the Langhe planar landslides is well known in Italy, less abroad. In effect, typing in Google Scholar "rock-block slides" 382 links may be found; but if we wanted to refine the search by including the "marly silty" and "arenaceous-sandy" voices, the search would provide only five published papers.

The Langhe Hills, particularly renowned for the cultivation of fine vineyards, are bordered by the Tanaro River to the north and west, the Orba River to the east, and the Apennines to the south [1]. Langhe covers an area of about 2300 km^2 in southeast Piedmont (northern Italy): the main city of the area is Alba, which lies on the banks of the Tanaro River. Alba, a town famous for the confectionery industry, is located about 150 km southwest from Milan and 65 km SSE from Turin and 70 km north-west from Genoa. Tanaro originates in the Apennines and flows north along the western boundary of the region until it reaches the small town of Cherasco, where, due to old capture phenomenon, it bends right and flows on to Alba and Asti, after which it joins the Po River downstream from Alessandria (**Figure 1**).

The Langhe is rolling hills famous for their excellent wine, populated by many small inhabited centers since ancient times: Alba is the largest and most important city. An idea of the Langhe geomorphology can be gained by studying the word "Langa": it may have been derived from either "landa," which means a wild and uninhabited place or from "lingua," which means a strip of land.

The geological characteristics of the layers, the geomorphology of the hills, the mineralogical composition of some levels, and the predisposition for translational movement mean that the Langhe hills can truly be considered a particular area that lends itself much to be studied.

During intense rainfall events, Langhe hills are usually affected by translational rock-block slides [2]. These are particular interesting in terms of:

- their typical attributes: the landslides occur in tertiary flyschoid complexes represented by a rhythmic series of marly silty and sandy-arenaceous rocks; sliding surfaces correspond to a thin marly clayey level, where infiltrating water deeply penetrates extensive systems of discontinuities;
- their widespread distribution during major rainfall events; and
- their high repetitiveness: some slopes have been affected several times in the last 100 years, the period over which reliable rainfall data are available.

2. Geological and geomorphological framework of the Langhe hills

The Langhe hills form the southern part of the Piedmont Tertiary basin, which is bounded by the Alps to the south and west, the Apennines to the east, and the Po Plain to the north. The basin's complex structural and stratigraphic history results from the post-Eocene phases of the Alpine orogeny and the opening of the north-western sector of the Mediterranean Sea [3]. The Langhe basin is composed of an Oligocene-Miocene sequence of terrigenous deposits that lie in unconformity above a pre-Cenozoic bedrock formed by Alpine and Apennine metamorphic units and characterized by a complex crustal geometry. The sedimentary sequence is more than 3500 m thick and is differentiated in time with respect to syn-sedimentary tectonics that are mainly associated with tensional forces [3]. Sedimentation begins with the deposition of continental deposits in alluvial fans and fan-deltas, followed by shallow marine transgressive deposits of the Late Eocene to Early Oligocene. The drowning of the area was marked by the deposition of open-sea units composed of alternating high-density flow deposits and hemipelagic sediments dating from the Early Oligocene to the Tortonian. The sedimentation ended in the Messinian with deposits that marked a sudden drop in sea level [4]. The main rock types are marls, silty marls, mudstones, sandstones, and conglomerates. The sequence forms a monocline that has a regional dip of 10–12° toward the northwest and is displaced at different stratigraphic levels by normal or strike-slip faults with a NE-SW orientation. Fault throws are usually in the range of 10–50 m, but locally they reach values of over 200 m [5].

The morphology of the Langhe hills is clearly affected by its structural setting. The area is characterized by the asymmetrical profile of the valleys, which results from isoclinal bedding in marly silty and arenaceous-sandy facies. Long, gentle slopes dip toward the northwest in the same direction as the bedding planes, although the inclination of the strata is usually lower than the slope surface. Short, steep slopes dip toward the southwest at about 25–40° in the opposite direction to the dip of the bedding. The morphological instability of the area is reflected in the common occurrence of landslides, which represent the main morphogenetic factors on the slopes. The gentler hillsides are prone to translational rock-block slides that involve the bedrock to a maximum depth of 20 m, while landslides due to saturation and fluidification of the eluvio-colluvial cover, the so-called "soil slips" [6] occur on the steeper slopes.

3. Past severe events and respective studies in the Langhe area

Since immemorial time, translational rock-block slides have been very common in the Langhe hills. In spite of this, in Italy, the scientific community started to show a certain interest in these phenomena only in the 1960s of the twentieth century. In the preceding period, the studies and surveys had been completely casual, despite the preeminent control role that these phenomena have always played on the morphological evolution of an extensive portion of the Piedmont territory and the reflexes for the safety of numerous inhabited centers, as the above chronology of **Table 1**. Some of the major movements of the past century are still remembered by old people in the Langhe villages. In an extensive analysis of instability, Forlati and Campus [7] identified over 2000 active and quiescent planar slides. Their morphostructural elements, which are mostly scarps and cracks, correspond clearly with those of old landslides. Govi and other authors [8–9] have demonstrated that, on many unstable slopes in the Langhe hills, instability along the bedding planes represents the same kind of evolution: in November 1994 meteorological event, for example, 44% of them were considered as a reactivation of a past movement [10].

The first historical events of which we have certain and documented information were two, those that at the end of March 1679 destroyed, respectively, the town of Bosia and part of the famous baths of the Acqui city [11]. They were two destructive TRBSs, which were mentioned in the local chronicles. Then for almost two centuries, there was no news, although it cannot be excluded that other episodes have happened in the meantime. But on the other hand, it is always difficult to find historical finds of the eighteenth century: since there are no large cities in the Langhe, but only small towns, the only documents that can be found are often those produced by the parish priests of the churches that reported the misfortunes that occurred in the villages.

The first events of the nineteenth century remembered by written documents are those that occurred in the period of 1853–1861 (three cases).

The first contributions of the twentieth century are due to the naturalist-geologist Sacco, who on 1901 and 1903 described the landslides of Mondovì [12] and Cherasco [13] advancing, for the latter, acute observations on the possible evolution and on the possibility of arresting the translational movement.

A few years later, De Alessandri published a very valuable study on the landslides of the Acqui area [14]. The author illustrated very carefully the historical planar slides of the Monte Stregone Mountain, underlining the tendency of the phenomena to reproduce in the same localities, noting that "it is an established principle that the landslides occurred are in their turn the main cause of further landslides and that their quiescence is often temporary, with intervals that can extend to a century, before resuming the movement phase."

During an intense rainfall period on May 16, 1926, a translational landslide occurred near the village of Levice [15], involving a surface area of 3 ha (**Figure 2**). According to eyewitness accounts, long cracks appeared less than an hour before large fractures 20–25 m in depth opened in the slope. There were no victims, and no buildings were destroyed, but meadows and vineyards were damaged.

Ten years later, on March 6, 1936, in the municipality of Castino, the small village of Vernetta was comprehensively damaged by a rock-block slide with a surface area of 35 ha. The blocks moved 150 m at an average velocity of 60 cm/h: a stretch of highway and seven houses were destroyed, and 38 people were evacuated [16].

On April 7, 1941, following a rainy period after an especially snowy winter, a sudden translational landslide destroyed the Cascina Bric farmhouse, near the village of Cissone: the mass movement killed three people and destroyed several buildings. The width of the main scarp was about 300 m, and the rock blocks were about 30 m deep (**Figure 3**). The phenomenon was carefully studied by Boni [17], with a good report accompanied by excellent photographs. The author highlighted the conditioning role, for the purposes of stability, of the geological-structural structure of the region and of the lithological characteristics of the rocks involved;

Year/mm/dd	Municipality/place	Brief description	Source
1679/03/31	Bosia	A large translational landslide devastated much of the ancient town. The landslide will be reactivated again in 1853 and more recently in November 1994.	[11, 18]
1679/03/31	Acqui	An "extraordinarily voluminous" landslide, slipped from Monte Stregone Mountain, buried the thermal establishments of Acqui Terme.	[14]
1860 approx.	Ciglié	"Half a hill collapsed, swallowing several houses in the disrupted clods."	[15]
1861/10/??	Rocchetta Belbo	A landslide caused "an unforgettable catastrophe."	[15]
1876/05/23	Mango	A rock-block slide of about 40 ha shifted a house for 120 m. The moved clods had a depth of up to 13 m.	[11]
1876/spring	Acqui	From the slope of Monte Stregone Mountain another great landslide broke off, adjacent to that of 1679, which destroyed houses, claiming human victims (number unknown).	[14]
1879/05/26	Igliano	Some witnesses observed the opening of a large fracture close to the church and then the sliding of an area of 2 ha for 15–20 m, with disarticulation in clods 15–20 m high. They were warned "a huge roar and a wobble of soil," and "flames, accompanied by a sulfurous smell" were observed.	[11, 19]
1892/03/31	Roccaverano	"Large planar landslide."	[12]
1892/03/31	Vesime	One rock-block slide moved 60 m together with a house causing two victims.	[19]
1901/03/20	Trezzo Tinella	A landslide with an area of several hectares swept over a farmhouse.	[11]
1901/03/20	Mondovì	A slippage about 20 m deep originated with a "general cracking" of the mass (probably over 200,000 m^3).	[11, 12]
1902/spring	Acqui area	Period sadly remembered for the large number of landslides. On the Monte Stregone Mountain, great dislocations were reported with "a fracturing and an overlapping of layers."	[13, 14]
1902/spring	Melazzo	Two houses were overwhelmed by a planar landslide.	[13]
1902/spring	Cherasco	A translational block slide with a width of over 300 m and a length of about 700 m involved the slope on which stood a cluster of farmhouses causing serious damage to the buildings.	[13]

Year/mm/dd	Municipality/place	Brief description	Source
1905/05/15	Lussito	Exceptionally abundant rainfall triggered numerous landslides. On the southern slope of Monte Stregone Mountain, there was "a slippage of a large fractured mass ... on a marly surface ... made extremely slippery."	[14]
1907/04/06	Acqui	Another landslide on the Monte Stregone Mountain that devastated an area adjacent to that collapsed in 1876. The mass involved (area 0.35 ha, volume 30,000 m^3) slipped on a marly stratification plane that "after the phenomenon had a highly laminated mirror surface." There were five casualties.	[14]
1917/05/29–31	Vesime	A 3-ha landslide swept over a building.	[19]
1917/05/29–31	Terzo	Another landslide removed a road for a length of about 80 m.	[19]
1926/05/16	Levice	During the severe alluvial event that hit the Tanaro and Bormida river basins, a landslide extended for about 3 ha, 10–15 m deep, marginally involved the Levice little town.	[15]
1926/05/16	Benevello	A "massive landslide" was reported.	[11]
1936/03/05–06	Castino	Sudden landslide of about 35 ha with damage to buildings, while the state road was moved by 150 m. The landslide will be reactivated later in 1953, but it moved only a few meters.	[11]
1941/04/07	Cissone	A small hamlet was moved about 60 m and destroyed (three casualties) by a sudden 30 m deep slide and about 300 m wide in the detachment area.	[17]
1951/02/10–12	Zone between the Belbo and Uzzone streams close Levice	An alluvial event hit the area between the Belbo and Uzzone torrents with particular gravity, causing "a very large number of landslides, subverting in many points the age-old stability of extensive cultivated and inhabited slopes. There are numerous cases in which entire vineyards ... they literally glided down the valley for tens and hundreds of meters, leaving theirs place a shiny reflecting surface."	[19]
1956/03/26–28	Niella Belbo	A translational landslide extended over 4 ha cut down two farmhouses.	[11]
1957/04/07–13	Trezzo Tinella	Important reactivation of the 1901 landslide.	[11]

Year/mm/dd	Municipality/place	Brief description	Source
1963/01/05–06	Cigliè	Progressive aggravation of the conditions of stability of the great phenomenon of Cigliè (surface 25 ha, thickness 15–25 m, volume 3.5–4 million m^3) with destruction of 16 houses.	[15]
1968/11/01–04	Gottasecca, Prunetto, Niella Belbo, and Bossolasco	During this serious event, numerous important planar landslides were activated, recognizable on aerial photographs of 1969 or documented by photographs from the ground. The largest was that of Bossolasco slide, extended 15 ha.	[11, 20, 21]
1971/ April–May	Castino	An extensive planar landslide involved some rural buildings.	[11]
1972/02–03 and 1974/03	Somano and Cherasco	These two events, which took place into a period of only 24 months, activated or reactivated numerous planar landslides in the Belbo, Tanaro, and Bormida valleys. We recall the great slide of Somano (10 million m^3) and that of Cherasco (4–5 million m^3). The 1972 event, with over 130 identified cases, associated with that of 1974, which reactivated numerous landslides that had arisen 2 years earlier, can be considered in terms of diffusion of phenomena, the second-greatest event in the Langhe after the November 1994.	[8, 11, 20, 22–24]
1977/10/07	Ovada area	Numerous deep TRBSs with an area of up to 1 ha originated between the Ovada area and the Scrivia River basin.	[19]
1985/03/08	Ovada area	One large rock-block slide involved a farm.	[20]
1994/02/06	Monastero Bormida	Planar landslide destroyed two buildings and part of an aqueduct.	[20, 25]
1994/11/05–06	Several dozen municipalities	A very intense and prolonged rainfall event produced more than 450 TRBSs in the Langhe hills. In the Ceva territory up to 10 TRBSs occurred in 1 km^2.	[18]
2019/11/26	Bubbio	Area involved equal to 1 hectare; translation of the clods about 20 m with maximum 7 m in height. Significant damage to the vineyards and the municipal road (removed 100 m).	Personal communication
2019/11/26	Santo Stefano Belbo	Landslide triggered in a vineyard. Rejection of only 70–80 cm, but sufficient to severely damage a farmhouse.	Personal communication

The words in the "quotation marks" are original of the reference document.

Table 1.
Most famous past translational rock-block slides in the Langhe hills for which at least one report has been found. These landslides are historically documented since the second half of the seventeenth century: in this chronology, briefly reported, the most significant cases for dimensions and effects produced are reported.

Figure 2.
May 16, 1926: Levice rock-block slides, top view. The large disjointed clods are evident.

Figure 3.
Cissone, Cascina Bric farmhouse. The large rock-block slide that caused three casualties on April 7, 1941.

he reconstructed the phases preceding the collapse and put forward some hypotheses on the determining causes of the landslide, recognizing and analyzing the triggering role of the rain that had fallen in the previous days and the preparatory role of the melting waters of the abundant snowfalls of the winter.

On January 9, 1963, the Cigliè rock-block slide destroyed 10 buildings, and 17 others had to be evacuated. The landslide had been known since the late eighteenth century and was reactivated in 1860, in the early twentieth century, and in 1954, 1956, and 1958, but the movements that occurred during the spring of 1960 were somewhat larger than previous ones. The total surface involved was about 25 ha, with a depth of 15–25 m and a volume of 4 million m^3 (**Figure 4**). The approach of Cortemiglia and Terranova [26] to this TRBS was mainly morphological-geological with marginal references to the role of rains. In the work, some data related to the geotechnical characterization of the rock mass were also exposed.

During the first week of November 1968, intense rainfall hit the Langhe area: many important planar landslides were activated, recognizable on aerial photographs of 1969 or documented by photographs from the ground. The largest was that of Bossolasco slide, extended 15 ha. Grasso [20] in a study dedicated to the Belbo Valley seriously affected by extensive flooding and numerous landslides in the aforesaid period, defined "landslides by plasticity" or "plastic-gravitational deformations in the eluvial cover located on the marly silty substrate" phenomena that, from what can be observed from the published photographs, they were presumably planar slides in the incipient stage. Only in the chapter, "geomorphological observations" were made an explicit but very brief reference to these landslides, which are considered "generally of limited extension."

In February and March 1972 and February 1974, several translational landslides occurred in the Langhe hills [8, 22]. Some of the movements affected entire slopes and caused severe damage to rural buildings, main highways, secondary roads, and agricultural activities (in particular vineyards and wineries). Two of these landslides, the Somano and Arnulfi rock-block slides, were extensively studied. The Somano landslide occurred on March 13, 1972 and was reactivated on February 18, 1974. With an area of almost 1 km^2 and a volume of about 10 million m^3, it was the largest rock-block slide to occur in the Langhe in the twentieth century (**Figure 5**).

Figure 4.
Detail of the landslide phenomenon of January 1963 in the crown area, in a photograph of the time. There are clearly visible bedrock structure dip slope of tertiary-age sedimentary rocks (on the left) and part of a slipped clod (on the right).

Figure 5.
Aerial photograph of the huge translational rock-block slide at Somano (February 1974). The movement stopped at the opposite slope causing a dam and consequent small lake of the Gamba stream.

Of more than 200 slopes identified as having been subject to translational movements, in 1972, some 130 showed clear signs of instability, although a smaller number manifested similar signs in 1974. The Arnulfi rock-block slide had a volume of about 5 million m^3 and affected a slope that showed signs of past instability.

The large number of TRBSs altogether triggered or reactivated by two pluviometric events close together over time (February to March 1972 and February 1974) gave inspiration to some researchers from the CNR-IRPI of Turin to focus attention on these phenomena. The first planar landslide studied in order of time was that of Altavilla di Somano of which Govi [8, 23] reconstructed with extreme accuracy the evolution on a multitemporal photointerpretation basis, while Sorzana [22], with reference to the Arnulfi landslide near Cherasco, underlined the role of hydrological-climatic parameters in play, between which the evaporation-transpiration.

The studies of Biancotti on the geomorphology of the Langhe [24, 27–29] highlighted the important morphogenetic role of landslides on dip slopes, for which models of dynamic evolution were illustrated by the author. Particular attention to these landslides was dedicated in the study on the Rea Stream basin, which also included a map of slope dynamics "which reported some slides (e.g., Somano, Cissone) and indicated some sectors of dip slopes prone to 'creeping and soliflux.'"

The investigations jointly undertaken by the CNR-IRPI of Turin together with the Geological Service of the Regione Piemonte in the years 1978–1980 for creating a systematic cartography of the Piedmont instability phenomena were another opportunity to expand and deepen the state of knowledge on sliding landslides. For the first time, the detailed picture of these phenomena was outlined, the actual areal distribution and the state of activity indicated with appropriate symbologies on maps at the scale of 1:100,000. The result of a decade of research carried out by the CNR-IRPI of Turin on the subject materialized into two summary papers:

- the first by Govi and Sorzana [8], valuable for its methodological approach, pointed out above all the kinematic-evolutionary aspects, highlighting how most of the slides analyzed had occurred on the slopes that had already been the site of similar phenomena in the past and

- the second by Govi et al. [9] identified the minimum cumulative rainfall threshold of an event lasting from 1 to 3 days capable of triggering a slide, in 100 mm, provided that the rainfall in the previous 60 days exceeded variable values from 150 mm to 300 m, depending on the season in which slides occur (usually between November and May).

In 1993, the book "Atlante dei Centri Instabili piemontesi" ("Atlas of unstable Piedmontese inhabited centers") was published by Luino et al. [15], in which some detailed illustrative sheets of important landslides appeared of sliding.

Then, in 1994, the most important event in the history of the Langhe hills took place: it was defined by experts as "secular." The event was devastating due to the flooding of the valley bottoms heavily urbanized causing enormous damage and 44 victims along the Tanaro Valley. But landslides on the slopes also made a notable contribution in terms of destroyed roads, damaged houses, and removed aqueducts. Many countries were totally isolated, and there were great problems in order to restore a certain normalcy.

On 5–6 November, after days in which rainfall reached maximum intensities of 50 mm/h and 200–250 mm/day, more than 450 translational landslides of various size occurred in the Langhe slopes (**Figure 6**) with a top of 10 rock-block slides within only 1 km^2 in an area around the Ceva town [18]. In terms of areas involved by various landslide typologies, the November 1994 event was similar to the February 1972 and February 1974 events.

In the years immediately following the serious event of November 1994, several authors addressed the study of these characteristic landslides, examining different aspects, in particular geology, morphology, hydrogeology, and predisposing and triggering factors. We remember Del Monaco et al. [30], Aleotti et al. [31], Polloni et al. [32], and Ayala et al. [33], who described some rock-block slides occurred during the event into a synthesis report of the field mission of a CEC research group carried out in the Langhe within the framework of the EU project MeFISSt.

A photointerpretation work was instead undertaken by Susella [34] on a large landslide that occurred not in 1994, but many centuries earlier, in a period difficult to define. Using of aerial photographs and field surveys, he identified the largest movement known to date, which occurred near the town of Cravanzana. Here, a landslide moved blocks 40–45 m in depth, over 2 km wide and 700 m long. The estimated volume exceeded 10 million cubic meters.

Again referring to the landslides of November 1994, the physical-mechanical parameters of TRBSs were studied by Lancellotta and Scavia [35], while in the 2-year period of 1997–1998, Aleotti together with other authors addressed the issue of TRBSs with two papers, which, starting from the November 1994 event, examined geological, structural, morphological, and meteorological conditions of the Langhe hills. In the first paper [36], the authors demonstrated that the cause of the original process is flexural slip between layers of different competencies. Murazzano rock-block slide was deeply analyzed and confirmed the validity of this hypothesis showing the possibility that large flexural slip displacement could have occurred in a thick incompetent unbonded sandy silt layer. In a second paper [37], the authors suggested that, based on the results of studies performed to point out the location of the prone areas, the typology of the expected phenomena, and their frequency, adequate hazard management policies could have been

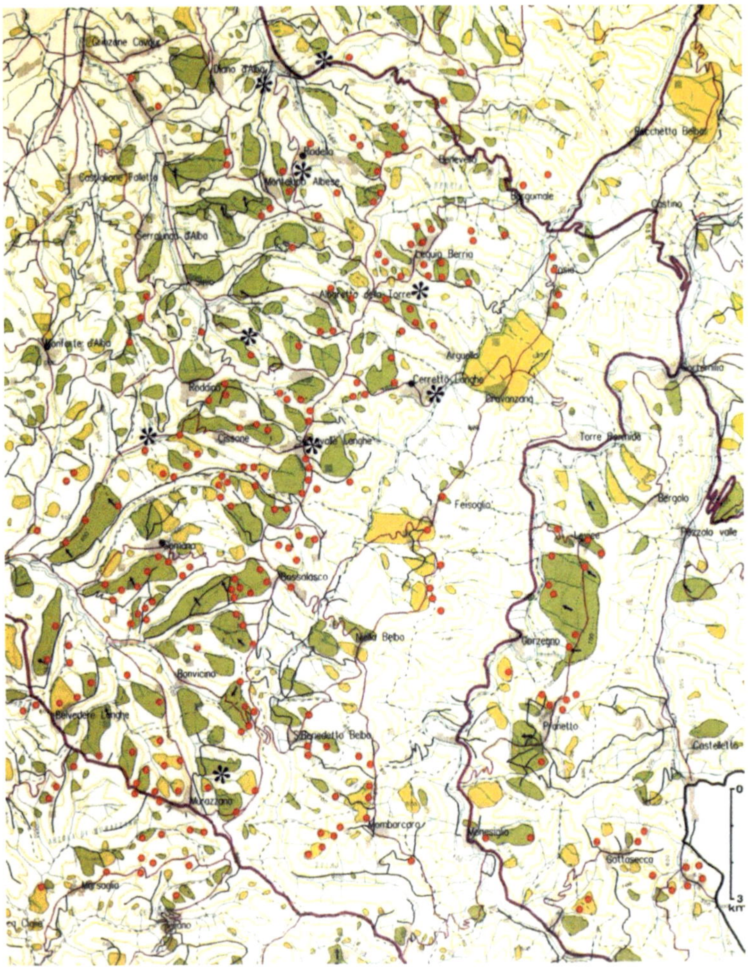

Figure 6.
Location of the main landslides of November 1994 in the Langhe area reported on the map of landslides (by CNR-IRPI and Regione Piemonte). The red spots indicate the TRBSs, and the blue asterisks indicate the areas in which there was the largest concentration of shallow landslides (soil slips). On the map, the quiescent TRBSs are indicated in yellow color; landslides active in the previous 30–40 years in green color (the small blue arrows indicate the most active sectors). In purple color, the roads are indicated.

undertaken especially in hilly area. They underlined that prevention or stabilizing measures are not always effective and in any case involve high costs; they concluded that for the elements that were already exposed at risk, monitoring and warning system with data teletransmission and real-time elaboration could have been a valid solution.

On 1998, Lollino et al. [38] sought to identify some factors capable of triggering and accelerating planar phenomena with the aim of a more in-depth understanding of the kinematic mechanisms and any related risks.

The TRBSs of the Langhe hills in the second half of the 1990s became the main topic study of the Regione Piemonte Geological Service. Its geologists and engineers deeply analyzed the geological, kinematic, and physical characteristics of the landslides: the papers of Forlati et al. [25, 39], Forlati et al. [40], Campus [41], Forlati and Tamberlani [42], Susella [43], and Forlati and Campus [7] are noteworthy. In this period, the experts of the Geological Service produced in particular: (1) cartographies of active and quiescent landslides (in collaboration with the CNR-IRPI of Turin, see **Figure 7**), (2) monographs of individual landslides, and (3) research on the geotechnical characterization of the slopes.

Figure 7.
Sliding surface of the Monastero Bormida landslide (February 6, 1994): The following days, the glossy surface was extremely slippery. It created great difficulties for geologists during the survey.

Other studies on rock-block slides of the Langhe hills were produced by Heiland and Stemberk [44], with an interesting comparison between these Piedmontese slides and similar phenomena in Czech Republic (Beskydy Mountains), Aiassa et al. [45], and Bandis et al. [46]. In 1999, Luino [18] presented the results of 4 years of studies on the Langhe and surveys carried out as a result of the aforementioned event in 1994, describing the triggering rains and the processes that occurred on the ground, both in the hydrographic network and on the slopes.

Mason and Rosembaum [47] used Landsat Thematic Mapper and multitemporal SPOT-Panchromatic image data to identify slope instabilities generated by the storm that struck Piedmont in November 1994. Slope angle and slope aspect data were derived from a Digital Elevation Model (DEM) produced from stereo air photographs. The geohazard map was then compiled by merging digital slope data with geotechnical characteristics, utilizing map algebra within a Geographical Information Systems (GISs).

Mandrone with some colleagues, in the period of 2000–2004 [48–50], assessed the risk conditions of the historic centers of the Langhe by analyzing the influence of precipitation on the triggering mechanisms of the instabilities, presenting a forecast model for the triggering of planar sliding landslides based on monitoring of the aquifer level and its correlation with meteorological data.

In the period of 2002–2004, with the aim to map all the existing landslides in Piedmont (including both results of monitoring data and available historical data), the Geological Service of Regione Piemonte participated at the so-called IFFI national project (Inventario dei Fenomeni Franosi in Italia – Inventory of Landslides in Italy) [51]. This landslide inventory represents a fundamental base of knowledge, is a very basic tool for land planning, and strongly helps the local authorities in their decision making. At the same time, the Agency for Environmental Protection of Regione Piemonte (ARPA) carried out new systematic surveys using aerial photo interpretation and created a specific alphanumeric GIS-based database to store and process all the collected data.

In 2005, Luino [52] published an article in which he analyzed the sequence of instability processes triggered by heavy rainfall with a particular focus about the rock-block slide of the Langhe hills triggered during the 1994 event. Luino analyzed the different periods of the event putting in evidence the highest concentration of mass movement trigger moment.

In 2013, Tiranti et al. [53] underlined the importance to establish a warning system capable of providing announcement of activation of the TRBSs with sufficient advance. For this reason, ARPA Piemonte developed a precipitation-threshold-based model setup on an extensive collection of historical data about the landslide movements (since 1917) and the related complete meteorological dataset. The output model can be tested by observations derived by the regional landslide monitoring network consisting of inclinometers and groundwater gauges managed by ARPA Piemonte.

TRBSs were also tackled by Notti et al. [54]: they studied very slow large landslides and the new generation of Persistent Scatterer PSI that allows to increase the density and the time series quality of interferometric data. They analyzed landslides belonging to different geological, geomorphologic, and land-use contexts and with different monitoring systems, in Western and Ligurian Alps, Langhe hills, and a portion of Northern Apennines.

4. Main physical characteristics of the translational rock-block slides

Throughout Europe, flyschoid rocks are commonly unstable. In Italy, the highest concentration can be found in the Langhe hills, which is the reason, in particular after 1994, it has become one of the areas most studied by geomorphologists. As they involve large sectors of slope and completely alter the previous morphological shape, translational rock-block slides are recognized as the phenomena that cause the greatest degree of displacement. We can briefly analyze the most important characteristics of these landslides.

4.1 Gradient of the slipping plane

TRBSs develop along bedding planes on slopes with gradients that vary from 8 to 15°, with 11° as the most common angle (**Figure** 7). They usually correspond to joints that form the interface between sandy-arenaceous and marly silty levels. The dominant direction of translation is nearly always parallel with the dip of the strata, even when the dip direction of the slope face is different. In the latter case, landslides develop at right angles to the slope. At the end of the movement, the sliding surfaces appear as a smooth, inclined plane that sometimes shows the shallow tracks left by the sliding rock block (**Figure** 8).

Sliding surfaces correspond to a thin marly clayey level, where infiltrating water deeply penetrates along systems of discontinuities. Forlati and other authors have discovered that the planar instability phenomena can be ascribed to swelling and to the mineralogy of the material involved. Laboratory testing of specimens immersed in water has revealed that cracks open parallel to the sliding surface. Diffractometric and mineralogical analyses performed on the same samples have shown that in such cracks, the smectite content is similar to the one of the sliding surface as measured in the field. In addition, scanning electron microscope revealed the importance of the fabric, and hence of interparticle links, in the failure mechanism [25]. Moreover, Lollino and Lollino [55], analyzing the landslide of Somano, claimed that the swelling pressure that develops on the sliding surface because of the clay mineral content is a factor that enables the blocks to slide for remarkable

Figure 8.
Sliding surface of the Murazzano landslide taken from uphill: The shallow tracks left by the sliding rock blocks are evident.

distances. These results suggest that the presence of smectite is a decisive factor in predisposing block for sliding that then occurs when there is a sudden increase in rainfall and ensuing hydration processes.

4.2 Area of the slope involved in the landslide

Rock-block slides can extend from a few tens to several thousand square meters. In fact, 46.6% of the landslides studied vary in area from 0.16 to 1.28 ha [5]. Many other rock-block slides triggered in the past are memorable as a result of their dimensions and their consequences for human settlements and activities (**Figure 9**). The Cigliè landslide (1963) had an area of about 25 ha, while Castino (1936) rock-block slide reached 35 ha. Somano landslide on March 1972 reached an area of almost 100 ha [23].

4.3 Thickness and volumes

About 58.3% of the landslides detected have thicknesses <5 m, 30.1% is between 5 and 10 m, while it significantly decreases for those between 10 and 15 m (10.3%)

Figure 9.
Murazzano Cascina Fascinea farmhouse, one of the most evident landslides of the 1994 event with an area of about 7.4 ha. The clods detached and moved for over 80 m had a height of 8–10 m.

and further decreases for landslides with thicknesses greater than 15 m (1.5%) [10]. In this last category, we can count the landslide of San Benedetto Belbo (**Figure 10**), which was the maximum known depth of any instability that occurred in November 1994 event.

The resulting volumes vary considerably, depending on the areal extent of movements and the thickness of bedrock involved. Volumes are estimated to range from a few hundreds to millions of cubic meters. For example, the Somano landslide in the period of 1972–1974 moved a rock-block slide of a volume of about 10 million m^3.

4.4 Velocity

During the main phase of the movements, obviously, instrumental data are missing; however, the information obtained on the spot and the few bibliographic data available agree in affirming the suddenness of events (developed and exhausted in a span of time included between a few minutes and a few hours). During the paroxysmal phase, the movement can take place at speeds ranging from 0.5 m/h (less than 1 cm/min, Vernetta landslide) to a few hundred m/h (about 5–6 m/min, landslides of Somano and San Benedetto Belbo). The landslide of Somano in 1972 moved involving a farmhouse: in spite of its speed, during the sliding of the rocky mass, the inhabitants were at the table for dinner. The building, in solidarity with the rocky clod, moved gently, and the householders barely noticed the movement only hearing the clink of glasses in the cupboard.

4.5 Displacements

In the wide range of cases recorded over the years, there have been landslides that have been translated by a few meters, but most of them by several tens of meters (**Figure 11**). The systematic analysis of the summary sheets and the results

Figure 10.
San Benedetto Belbo rock-block slide. This landslide is the one with the maximum known thickness: The clods in some points were 22–23 m high, a value that can be deduced from the presence of the geologist on the top. During the late evening of November 5, 1994, eyewitnesses interviewed in the following days said that on the slope of the landslide, they saw many flashes originating from the rubbing of the rock on the sliding surface.

of the aerial photography interpretation analysis made it possible to classify the planar slides typologically, considering the geometry of the sliding surface and the geometry of the landslide body. Forlati and Campus [7] correlated the thickness of the shifted masses and the different types of landslide noting that the thicknesses most at play reached up to 4 m in the case of open plane (70% of the cases) and free over-current (65%) slips in free evolution, while in the case of confined evolution, they increased to 6 m (70%). They are therefore landslides that, although involving the substrate, do not deepen much, remaining on average values of thickness of mobilized clods within 6 m.

All types of sliding are therefore joined by rather modest thicknesses, without indicating a predominant trend compared to the others. An examination of Forlati and Campus [7] summary sheets yielded a population of 108 measures useful for assessing elongation, that is, the distance between the main slope and the lower limit of the accumulation. **Figure 12** shows that 65% of the measurements correspond to an elongation within 100 m. The passage from the discrete to the continuous representation was carried out by means of an exponential function that interpolates the histogram with a correlation coefficient equal to 0.92, thus obtaining a first level instrument for the evaluation of the zones of influence of the translated masses.

Figure 11.
Rock-block slide near Feisoglio town: The mass movement slid about 50–60 m from their original position, destroying a road. The estimated area was about 4.63 ha, while the volume was about 385,000 m³. The gray layer surface is evident. The small white house moved together with the rock block nearly undamaged.

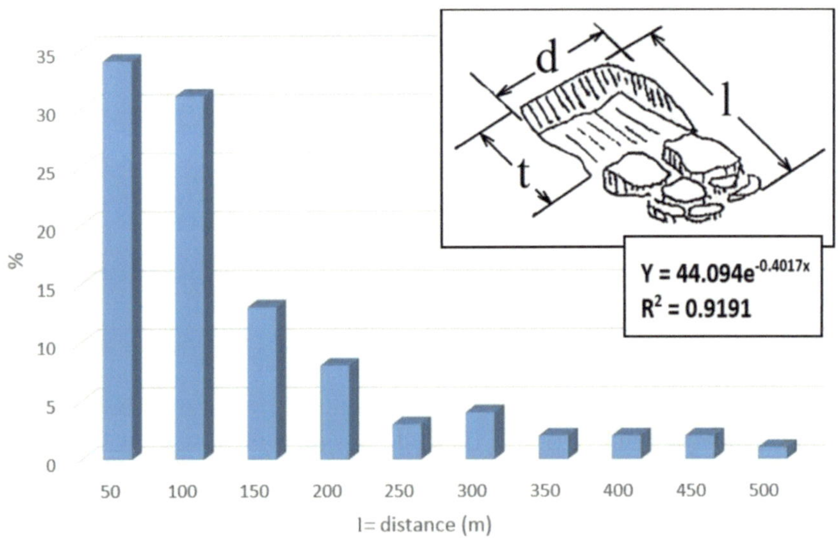

Figure 12.
Distance between main scarp and lower edge of displaced mass. Over 65% of the distances are <100 m according to an analysis of 108 TRBSs in the Langhe hills in the last 25 years ([7], modified).

5. Phases of rock-block sliding movements

In relation to conditions of progressive slope destabilization, studies of past rock-block slides [7, 8, 30] identified four phases of development.

In the first phase, which can continue also for some decades, sub-rectilinear cracks and tension gashes in the ground surface open and they develop to considerable depths. They are generally discontinuous or isolated events and occur primarily in the upper part of the slope. Concurrently, a swelling usually occurs in the middle and lower parts of the slope. In some cases, this phase is characterized by very limited premonitory signs that occur only a few hours or minutes before the sliding.

In cases in which the dynamics of sliding has been recognized, researchers have noted that the time interval between the appearance of the first premonitory signs and the sliding have not exceed 4 h in nearly three quarters of cases ([5], modified).

The second phase (**Figure 13**) differs from the previous one not for the typology of the processes but for the persistence in time of the described disasters that are repeated for the most part in the same places even when they are subject to readjustments for anthropic intervention.

The general conditions remain on average apparently stationary for many years, and the only indication of a growing instability is sometimes provided by the appearance of new small fissures in the ground, arranged in groups, soon, however, partially remodeled by the agro-cultural activities.

In the third phase, a portion of the slope collapses, usually without warning, resulting in the sliding of huge, disjointed rock blocks. The movements can take place over periods that vary between a few minutes for small landslides and some hours for large ones, and it may develop as a result of the erosion of the main scarp. At a later stage, the crown of the landslide may retreat, and the accumulated deposit may become partially fluidified. The mobilized masses slide along stratigraphic surfaces and often preserve the characteristics of the bedrock, even with respect to bedding, particularly over relatively short distances. In some cases, buildings have been propelled forward together with the rock block without being appreciably damaged and showing only cracks in their walls.

The last phase of the movement leads to new conditions of equilibrium, which are achieved by a slow succession of adapting movements that develop with

Figure 13.
Spectacular example of rock-block slide in a second stage characterized by a period of quiescence that can last for several years. During this time, the density of fracture systems increases greatly, leading to vertical displacements and the formation of morphological steps. Cracks can be some hundreds of meters long, often coinciding with the perimetrical fractures of older translational block slides.

progressive demolition of the blocks as a result of natural or human activity. Within a relatively short period of time, slopes are remodeled by the formation of a continuous colluvial cover that tends to hide the characteristic elements of the original landslides.

6. The role of rainfall

The majority of rock-block slides occur in concomitance with long periods of rainfall, even when these are of only moderate intensity. Govi et al. [9] studied the planar landslides that occurred before 1975 and identified a relationship between hourly rainfall intensity and critical rainfall, in which the effective rainfall of a single event is expressed as a percentage of the mean annual rainfall (**Figure 14**). The November 1994 event confirmed the results obtained by these authors. They analyzed the total rainfall delivered by the triggering event, antecedent precipitation (up to 60 days before the event), and the monthly distribution of rock-block slides.

Slope instability usually occurred after periods of heavy rainfall. The prior minimum rainfall needed to reach critical soil conditions varied according to the month: they were the highest in the period preceding the November event. Conditions for instability involved a combination of 60 rainy days and high-intensity precipitation. The rain that fell during the whole 2-month period fed the permeable horizons in the stratigraphic sequence to a critical level. The precipitation mainly influenced the volumes of stored water along fractures and rock discontinuities. The actions of the two variables may not cause instability separately, but together, they can trigger a rock-block slide.

Hourly intensities play an important role in the identification of the different phases. **Figure 15** shows the times of onset and collapse of the planar landslides plotted against hourly precipitation values. However, these data were collected for less than 25% of all movements that occurred in November 1994, and hence,

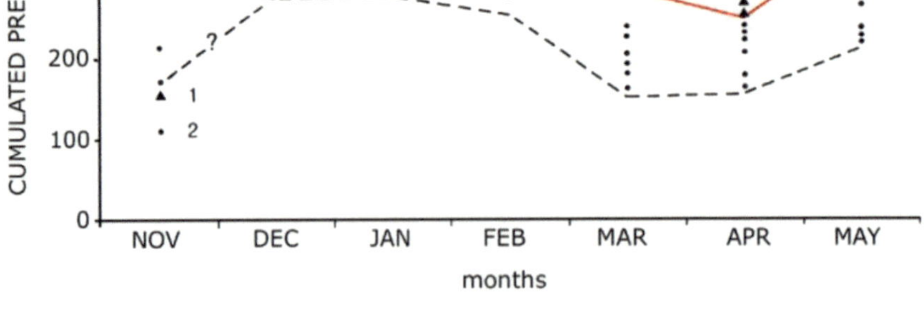

Figure 14.
Diagram showing precipitation amounts triggering rock-block slides in the Langhe hills. The dashed line defines the area of minimum rainfall values in the 60 days before the collapse. The red line denotes the threshold total values (previous rainfall + event rainfall) that cause landslides (1 – cumulated precipitations triggering landslides; 2 – cumulated rainfalls that do not trigger landslides). The rainfall values triggering the Somano and Rodello TRBS occurred on November 1994 are indicated with a red arrow ([9], modified).

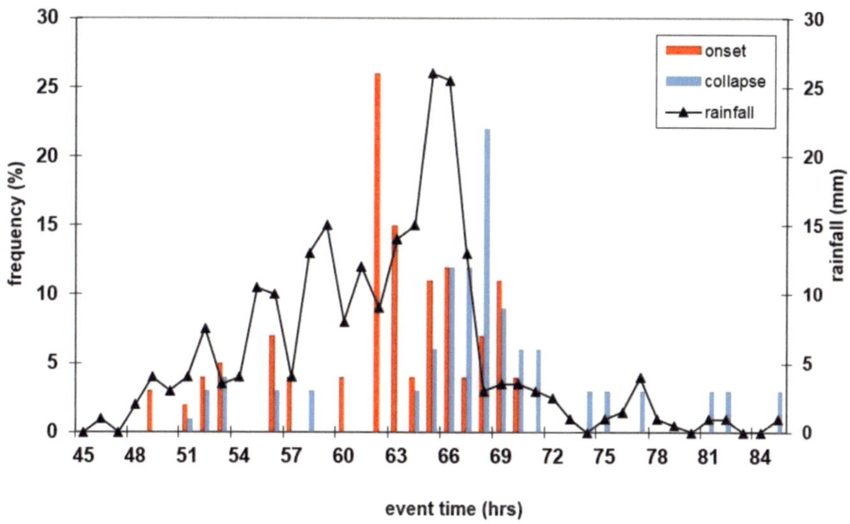

Figure 15.
Relationship between precipitation and time of onset and collapse of translational block slides in the Langhe hills during the severe November 1994 event.

the plotted values may not be representative of failure times as a whole, as it can be argued that the times of onset and collapse are more likely to be noticed during daylight hours than at night. The peak level of the storm was shown to have occurred at 65 h. The threshold total value was reached after 53 h. The climax of the failures occurred at 62 h, while the peak level of collapses occurred after 68 h of rainfall [36]. These observations indicate that there was a time lag of about 6 h between maximum rainfall intensity and failure.

The history of the major events highlights how TRBS always occurs for periods of intense rainfall that exceed a certain cumulative value.

For this reason, although we often hear about the role of climate change in progress, it seems difficult to find a good correlation between the frequency of the processes and the annual cumulated rainfall. In this regard, the graph in **Figure 16**

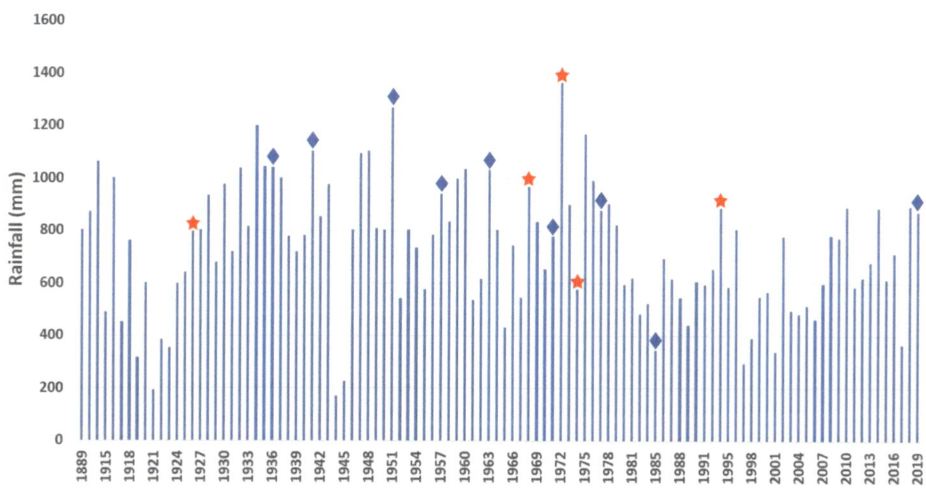

Figure 16.
Cumulative annual rainfall recorded in Alba (1889–2019 with some missing years), the most significant measurement station for the Langhe hills. The blue rhombuses indicate the years in which the minor TRBS occurred. The red stars indicate the years in which the major landslides took place.

shows the cumulative precipitation for each year since 1889 (with some gaps) for the Alba station. The symbols indicate the years in which the TRBS occurred (see **Table 1**).

Considering the seasons from a meteorological point of view, that is, winter from December 1 to the end of February, spring from March 1 to May 31, and so on, we can examine the 27 planar landslide events for which we know the month of occurrence (**Table 1**): it can be noted that as many as 18 of 27 cases (66%) occurred in the spring linked to the concomitance of the rainfall and the melting of the snowpack.

7. Damage

As previously described, in their movement, these landslides often involve structures and infrastructure. The percentage of buildings threatened is higher than that of destroyed or damaged buildings. In 1994, for example, if we take into account that many settlements were not directly affected by the landslides that took place, we can believe that there is a direct knowledge of the problem that led to a physiological selection over time of the safest areas to be used as residence. The same statement loses meaning for the road network where the relationship is reversed: that is, there are fewer cases in which the road network is threatened than those in which the road network is directly affected. Probably, this is due to the fact that the development of communications, as a need to maintain activities in this area, has undergone a significant increase compared to residential urban development which has been rather oriented toward the recovery of the existing one. Following the November 1994 event, more than 1300 TRBSs were recorded, of which almost half caused damage. Over a hundred landslides affected the road system, causing injuries for a total length of over 23 km, equal to about 230 m for each individual gravitational process [56]. The analysis of the damage caused to the inhabited centers by the action of the planar sliding landslides triggered during the November 1994 event highlighted how the highest percentage of damage was recorded for the houses located along the supper and middle sectors of the slopes. On the contrary, buildings built at the foot or near the crest have been less prone to damage in percentage terms. Examining the location of the historical urbanized areas, it is noted that, in the area most frequently affected by planar slides, 64.5% is located in a crest position and only 9.7% in the central sections of the slope [56]. The choice of the most suitable sites for the construction of the settlement centers therefore derives from the knowledge of the phenomenon of stability and the awareness of the effects by the population. The finding that the numerous planar landslides of November 1994 did not cause victims, as, moreover, observed almost generally also in other episodes of the past, suggests that in the population, there is a good historical memory of the process, which depends on the high frequency with which this type of landslide is proposed again and from the daily relationship with the territory, with a predominantly agricultural vocation.

8. Relationships between human activities and translational block slides

For the most of the landslides examined, the anthropic role has been very important: it occurred through actions that have modified the natural distribution of both surface and underground outflows. In this context, a predominant role is attributable to the significant development assumed by the road network, through new tracks and the restoration of old rural cart roads. Immediately following, in order of importance, the effects due to agricultural processes follow.

The analysis of the TRBSs of the Langhe hills mainly for the 1972, 1974, and 1994 events indicates that there is an evident correlation between the activation of movements and the human activities. While in many cases human-induced causes of landslides can be identified, the immediate triggering factor in this case is always an intense and prolonged rainfall. Human activity and land use both increase the potential for landslide activation, primarily where they modify the natural balance of surface runoff and subsurface infiltration.

It has been repeatedly observed that the slope cuts for the opening of service roads, exposing the cracked substrate and thus facilitating the infiltration of water into the subsoil are preferential places for the development of the first movement of land, with extensive opening of the cracks and subsequent propagation of the instability in the closest zones (**Figure 17**).

Similar considerations can be made with regard to the terracing and excavation works that interrupt the continuity of the layers, carried out to reduce the slope of the land for agricultural purposes or around residential settlements. The plows with deep furrows and the heavily engraved drainage network, almost always without protective channels, presumably constitute other important reasons, predisposing to the reactivation of the slipping.

But perhaps the most important aspect is the type of cultivation called "rittochino" or "franapoggio." The first is a hydraulic-agrarian arrangement of the sloping lands. The purpose of this arrangement is to regulate the flow of water while simultaneously reducing the risks of erosion and landslides. The lines on which the cultivation units and the hydraulic-agricultural products are developed follow the lines of maximum slope: the processes, the rows of arboreal plants, and the drainages develop orthogonally to the isohypses in order to favor rapid runoff of rainwater preventing excessive infiltration into the ground. It is probably the

Figure 17.
Murazzano rock-block slide, one of the most showy TRBSs triggered in November 1994, here shot a week later when a temporary road had already been restored. The clods moved along a surface with a gradient of 9.8° for a distance of about 80–100 m. The crown was located about 20–25 m upstream of the provincial road.

oldest settlement adopted in many hilly regions of Italy. The second, "franapoggio," instead, derives from the fact that the workings follow the isohypses: the cultivation is made by furrows arranged in a normal direction to the lines of maximum slope. The advent of agricultural mechanization has determined the almost total abandonment of the "reggipoggio" because tractors work in better conditions on the lines of maximum slope. This cultivation is still present in viticulture.

The most evident aspects of human activities that cause landslides in the Langhe hills are the following ones: (1) surface runoff is concentrated where roads are built in line with the fracture planes of past landslides; (2) water stagnates upstream of road embankments that lack drains; (3) the secondary hydrographical network is occluded due to agricultural work; and (4) water leaks from artificial reservoirs excavated on the slope in order to store rainwater [8].

9. Discussion

The fact is that the numerous rock-block slides did not cause many casualties, on November 1994 and also in other episodes of the past, as previously mentioned, it suggests that there is a certain historical memory of the problem, which depends on the high frequency with which this type of landslide is repeated and from the daily relationship with the territory, with a predominantly agricultural vocation. This entailed an accurate selection of areas to be used for permanent settlement purposes (the damage analysis shows that the percentage of buildings that are threatened is significantly higher than the destroyed ones), and the surprising sensitivity to perceive comes from the same reasons and distinguishes the warning signs of instability.

Thanks to this awareness of the problem and also bearing in mind that the preparatory phase of landslides due to gliding can last for several years and that the interval between the appearance of warning signs and the collapse is generally more than 2 h, it can be assumed that in the Langhe hills, an increase in the socially acceptable risk threshold has been reached, which must not, however, exempt us from deepening our knowledge of the problem. Indeed, the progress of knowledge confirms what has already been reported at the conclusion of the study cited on landslides in the Langhe of the Cuneo province in the years 1972–1974, and that is, the slopes in question need to be considered areas with high risk of landslide, even if there are not always perceptible signs of instability.

The knowledge gained on rock-block slides was further enriched with the alluvial event of November 2–6, 1994, which proposed a great variability of situations of great importance for the understanding of these phenomena: an extraordinary, and perhaps unrepeatable, "didactic gym," as well as a field of intervention for the competences of the Public Administration, which has provided several points for reflection, some of which are briefly outlined here. The significant morphological effects on the slopes and the spread of damage to the roads and the residential fabric led, a few days after the event, to the creation of a Regional Advisory Commission with tasks of methodological orientation and support to the operating bodies. Among the first products, as early as January 1995, a summary map was made available on a scale of 1:25,000 of the landslide phenomena recognized by photo-interpretative study, accompanied by an appropriate "data base" containing considerations on the evolutionary stage and damage caused by each of the 470 census landslides. The creation of a suitable survey card allowed us to collect data from 200 sample landslides on the ground. The detailed mapping on a scale of 1:10,000 and the subsequent digitalization also represented the justification for a complete review of the entire area and its evolutionary history.

The extensive amount of data acquired allowed a comparison between the events of the 2-year period of 1972–1974 (on a window of 200 km^2): as many as 130 landslides of 200 recognized, even as more ancient phenomena, were reactivated in that period [8]. Not different was the 1994 event, as 46% of the census landslides were considered as reactivations of previous phenomena.

For large sectors (e.g., area of Somano, Serravalle Langhe, and Bossolasco), it was possible to speak of "persistent intermediate stage" with episodic aggravations, which are also important, distinguishable over time.

A systematic photo-interpretative analysis of planar slips activated in spring 1972 compared with those of the autumn 1994 event allowed further significant considerations on the dynamic and evolutionary characteristics of these phenomena, namely:

- widespread reactivation of sectors or portions already place of movement in the 2-year period of 1972–1974;
- clear conditioning of the most significant morphostructural elements already manifested in that period;
- complete evolution of sectors in which signs of incipient instability were present; and
- reappearance on artificially reshaped slopes of fractures and traction joints that had affected the movements in 1972–1974.

The same considerations can be valid, in general terms, also for sectors of slopes that have been affected by landslides prior to 1972. It can therefore be argued that many of the areas affected by movements in 1972 were reactivated, especially for the larger phenomena, in 1994 with various spatial modalities but mainly using the same structural guidelines.

Based on testimonies collected from the affected population, an interesting information was provided regarding the perception of the warning signs and the times of landslide development. In general, it was found that for about 60% of the 70 cases analyzed, the interval between the appearance of the warning signs (cracks on the artifacts and in the ground) and the collapse phase is about 2–4 h.

10. Conclusions

The purpose of this paper is to make TRBSs known to the international scientific community, showing a synthesis of the studies and history of about these phenomena in the Langhe region, in north-western Italy. The Langhe delineates themselves, in the articulated panorama of the Piedmontese relief geological instability, due to the high presence and repetitiveness over time of the planar landslide process. Alongside important episodes of activations widespread in 1972, 1974, and 1994, still recognizable on the slopes, especially with the help of aerial photographs, there are signs of ancient movements, also large sizes.

These particular landslides clearly reveal the area's intrinsic morphological instability and the principal morphogenetic factors on slopes. Moreover, slope instability creates huge problems for the people of this hilly zone, not so much for the number of victims (<15 in the 340 years analyzed in **Table 1**), but for the extent of the land involved: just think that in some municipal areas such as Serravalle Langhe and Bossolasco, the area involved in landslides represents 40–42% of the total area [5].

During major hydrological events, main highways, secondary road networks, farmland, and rural buildings are often damaged. Field surveys can provide direct knowledge of the contemporary problem and its long history. Fortunately, unlike some earlier landslides, the more than 450 translational landslides that occurred in November 1994 claimed no casualties, but these are always possible, as the rock-block slides are potentially highly dangerous. Given the short-time lapse between maximum rainfall intensity and rock-block slide failure, an adequate meteorological network is needed with the capacity to detect thresholds of alarm during precipitation and snow-melt events, so that public authorities can be prewarned of emergency situations.

Acknowledgements

The authors would like to thank Dr. Giovanni Mortara for advice in drafting the paper, Dr. Luca Lanteri for specific data on the involvement of the Langhe roads, and Dr. Manuela Bassi for historical rainfall data. The authors would also like to thank Dr. Barbara Bono for **Figure 1**.

Author details

Fabio Luino* and Laura Turconi
Consiglio Nazionale delle Ricerche, Istituto di Ricerca per la Protezione Idrogeologica (CNR-IRPI), Torino, Italy

*Address all correspondence to: fabio.luino@irpi.cnr.it

References

[1] Giammarino S. Evoluzione delle Alpi Marittime liguri e sue relazioni con il Bacino terziario del Piemonte ed. il Mar Ligure. Atti dell Societa Toscano di Scienze Naturale, Memorie Serie A. 1984;**91**:155-179

[2] Varnes DJ. Slope movement types and processes, Landslides analysis and control. In: Schuster RL, Krizek RJ, editors. Transp. Research Nat. Sc. Spec. Report. Vol. 176. 1978. pp. 11-33

[3] D'Atri A, Dela Pierre F, Gelati R, Gnaccolini M, Piana F, Polino R. Il Bacino Terziario Piemontese. In: Polino R., coordinatore, "Il Sistema Alpino-Appenninico nel Cenozoico", Guida alle escursioni 6-9 Settembre 2002, 81a Riunione Estiva della Società Geologica Italiana "Cinematiche collisionali: tra esumazione e sedimentazione", 10-12 Settembre 2002. Turin, Italy; 2002. pp. 110-115

[4] Gelati R, Gnaccolini M. Genesis and evolution of the Langhe Basin, with emphasis on the latest Oligocene-Earliest Miocene and Serravallian. Atti Ticinensi di Scienze della Terra. 2003;**44**:3-18

[5] Campus S, Forlati F, Nicolo G, Fontan D, Sorzana P, Gelati R, et al. Note illustrative della Carta della pericolosità per instabilità dei versanti. Foglio 211: Dego, alla scala 1:50.000. Servizio Geologico d'Italia, Arpa Piemonte; 2005

[6] Campbell RH. Soil slip, debris flows and rainstorms in the Santa Monica Mountains and vicinity, Southern California. U.S. Geol. Survey, Prof. Paper 851; 1975, p. 51

[7] Forlati F, Campus S. Scivolamenti planari nelle Langhe piemontesi: individuazione, elaborazione ed. analisi di alcuni elementi significativi. In: Luino F, editor. "La previsione delle catastrofi idrogeologiche: il contributo della ricerca scientifica". Proceedings of the Conference "Alba 96", Alba (Italia) 5-7 novembre 1996; 1998, pp. 173-183

[8] Govi M, Sorzana PF. Frane di scivolamento nelle Langhe cuneesi: Febbraio-Marzo 1972, Febbraio 1974. Bollettino Associazione Mineraria Subalpina. 1982;**XIX, 231**(1-2):-264

[9] Govi M, Mortara G, Sorzana PF. Eventi idrologici e frane. Geologia Applicata e Idrogeologia. Bari. 1985;**20**(II):359-375

[10] Forlati F, Ramasco M, Susella G, Campus S. Gli scivolamenti planari nel territorio delle Langhe piemontesi. L'evento del 4-6 novembre 1994. Esame del quadro conoscitivo in funzione di analisi previsionali e di gestione territoriale. Convegno su "La stabilità del suolo in Italia: zonazione sismica-frane", Roma 30-31 maggio 1996, Accademia Nazionale dei Lincei. In: Atti dei Convegni Lincei. Roma. n. 134; 1997. pp. 139-144

[11] Tropeano D, Terzano P. Eventi alluvionali nel bacino del Belbo: tipologia e frequenza dei dissesti in base a notizie storiche. Bollettino dell'Associazione Mineraria Subalpina. 1987;**24**(3-4):437-474

[12] Sacco F. La frana di Mondovì. Annali Regia Accademia. di Agricoltura. 1901;**44**:6

[13] Sacco F. La frana di S. Antonio nel territorio di Cherasco. Annali Regia Accademia. di Agricoltura. 1903;**46**:8

[14] De Alessandri G. Le frane nei dintorni di Acqui. Atti Società Italiana di Scienze Naturali. 1907;**46**:58-72

[15] Luino F, Ramasco M, Susella G. Atlante dei centri abitati instabili piemontesi. Pubbl. GNDCI, n. 964,

programma SCAI. Tip. L'Artistica, Savigliano; 1993. p. 245

[16] Corpo Reale del Genio Civile di Cuneo: Relazione sul movimento franoso verificatosi in zona Vernetta del Comune di Castino; 1936. Report unpublished

[17] Boni A. Distacco e scivolamento di masse a Cissone, frazione di Serravalle delle Langhe. Geofisica Pura e Applicata. 1941;**3**:142. DOI: 10.1007/BF02102839

[18] Luino F. The Flood and Landslide Event of November 4: 61,994 in Piedmont Region (Northwestern Italy): Causes and Related Effects in Tanaro Valley. "XXII General Assembly of European Geophysical Society". Vienna (Austria). 21-25 April 1997. Ed. Elsevier Science Ltd., Vol. 24; 1999. pp. 123-129

[19] Tropeano D. Eventi alluvionali e frane nel bacino della Bormida. Studio retrospettivo. Quaderni di Studi e di Documentazione dell'Associazione Mineraria Subalpina. 1989;(10):155

[20] Grasso F.: Studi per la sistemazione idrogeologica della Valle Belbo. In CREEP ed.: L'agricoltura delle principali zone piemontesi colpite dalle alluvioni del novembre 1968. Sirea Print; 1969. pp. 237-278

[21] CNR IRPI of Turin: Photographic archive; 1970-2019

[22] Sorzana PF. La frana di Arnulfi nel Comune di Cherasco (CN) (febbraio 1974). Bollettino dell'Associazione Mineraria Subalpina. 1980;**XVII**(2):505-526

[23] Govi M. La frana di Somano (Langhe cuneesi). Studi Trentini di Scienze Naturali. 1974;**51**(2A):153-165

[24] Biancotti A, Di Maio M, Franceschetti B. Analisi ecologiche applicate alla difesa del Bacino del Torrente Rea. In: REGIONE PIEMONTE-Ass. Organizzazione Gestione Territorio, Studi dell'IRES sui piani di sistemazione idrogeologica; 1979. p. 332

[25] Forlati F, Lancellotta R, Osella A, Scavia C, Veniale F. Analisi dei fenomeni di scivolamento planare nelle Langhe, 213-217. GEAM n. 87, n. 4 "Difesa del Suolo"; 1995b

[26] Cortemiglia GC, Terranova G. La frana di Cigliè nelle Langhe. Società Geologica Italiana. 1969;**VIII**(2):145-153

[27] Biancotti A. Geomorfologia delle Langhe: il bacino del Fiume Bormida di Millesimo. Geografia Fisica e Dinamica Quaternaria. 1981a;**4**:87-101

[28] Biancotti A. Geomorfologia delle Langhe sud-occidentali. Memorie dell'Accademia delle Scienze di Torino, serie Classe di Scienze Fisiche, Matematiche e Naturali. 1981b;(5):1-21

[29] Biancotti A. Geomorfologia dell'Alta Langa (Piemonte Meridionale). Memorie Società Italiana di Scienze Naturali. 1981c;**22**(3):59-104

[30] Del Monaco G, Dutto F, Mortara G. Landslides. In: Casale R, Margottini C, editors. Meteorological Events and Natural Disasters: an appraisal of the Piedmont (North Italy) case history of 4-6 November 1994 by a CEC field mission. Arti Grafiche Tilligraf S.p.A., Roma; 1995. pp. 39-51

[31] Aleotti P, Baldelli P, Polloni G. Landsliding and flooding event triggered by heavy rains in the Tanaro basin (Italy). International Congress Interpraevent 1996: Changes within the natural and cultural habitat and its consequences, Tagespublication. band 1; 1996. pp. 435-446

[32] Polloni G, Aleotti P, Baldelli P, Nosetto A, Casavecchia K. Heavy rain triggered landslides in the Alba area during the November 1994 flooding

event in the Piemonte Region (Italy). In: Proc. 7th Int. Symp. On Landslides, Vol. 2. Trondheim: Balkema Publ.; 1996. pp. 721-725

[33] Ayala FJ, Bandis S, Del Monaco G, De Lotto P, D'Epifanio A, Dutto F, et al. In: Casale R, Margottini C, editors. Meteorological events and natural disasters: An appraisal of the Piedmont (North-Italy) case history of 4-6 November 1994 by a CEC field mission. Roma; 1996. p. 96

[34] Susella GF. Segnalazione di un antico e potente fenomeno di movimento di massa per scivolamento planare in Comune di Cravanzana nell'area delle Langhe piemontesi. GEAM, 2-3, giugno-settembre; 1996. pp. 73-78

[35] Lancellotta R, Scavia C. Frane per scivolamento planare delle Langhe: il contributo della ricerca all'analisi dei parametri fisico-meccanici. In: Le frane per scivolamento planare delle Langhe, Giornate di studio sui processi d'instabilità naturali, a cura della Struttura Studi e Ricerche-Banca Dati Geologica, Torino 20 ottobre 1997, Regione Piemonte; 1997. p. 6

[36] Aleotti P, Crosta G, Nosetto A. Diffused rock-block sliding: statistical, morphological and stability analyses. In: Proceedings of International Congress, IAEG, Athens; June 1997

[37] Aleotti P, Baldelli P, Polloni G. Soil slips, rock block slides and stream hydraulic processes caused by heavy rains: their interaction and relevant hazard. In: Proceeding of the Second International Conference on Environmental Management (ICEM2), 10-13 February 1998. Wollongong – Australia. pp. 553-564

[38] Lollino G, Lollino P, Bottino G. Analisi di stabilità e modellazione numerica di un fenomeno di scivolamento planare delle Langhe. In: Convegno Giovani Ricercatori, Chieti. ottobre 1998. pp. 1-10

[39] Forlati F, Brovero M, Campus S. Alcune considerazioni sulle deformazioni gravitative profonde di versante inerenti il territorio piemontese. Atti del II Incontro Internazionale dei Giovani Ricercatori in Geologia Applicata (I.M.Y.R.A.G.), Peveragno (CN) 11-13 ottobre 1995, Politecnico di Torino, Università di Torino, CNR-IRPI di Torino. Sez. A, 1995a. pp. 75-81

[40] Forlati F, Lancellotta R, Osella A, Scavia C, Veniale F. The role of swelling marl in planar slides in the Langhe region. In: Senneset K, editor. Landslides, Seventh International Symposium, Trondheim 17-21 giugno 1996, Balkema, Rotterdam, vol. 2; 1996. pp. 721-725

[41] Campus S. Elementi ed. aspetti salienti per l'analisi del processo d'instabilità naturale per scivolamento planare. In: Le frane per scivolamento planare delle Langhe, Giornate di studio sui processi di instabilità naturali, a cura della Struttura Studi e Ricerche-Banca Dati Geologica, Torino 20 ottobre 1997. Regione Piemonte; 1997. p. 23

[42] Forlati F, Tamberlani F. La Banca Dati Geotecnica: classificazione preliminare delle Marne Oligo-Mioceniche. In: Le frane per scivolamento planare delle Langhe, Giornate di Studio sui processi di instabilità naturali, a cura della Struttura Studi e Ricerche-Banca Dati Geologica, Torino 20 ottobre 1997. Regione Piemonte; 1997. p. 16

[43] Susella G. Scivolamenti planari nelle Langhe: Casi storici ed. esame del contesto geomorfologico. In: Le frane per scivolamento planare delle Langhe, Giornate di studio sui processi di instabilità naturali, a cura della Struttura Studi e Ricerche Banca Dati Geologica, Torino 20 ottobre 1997. Regione Piemonte; 1997. p. 15

[44] Heiland J, Stemberk J. Comparison of translational block-type slope movements in several flysch areas. In: Luino F, editor. "La previsione delle catastrofi idrogeologiche: il contributo della ricerca scientifica". Proceedings of the Conference "Alba 96", Alba (Italia) 5-7 novembre 1996, vol. 2; 1998. pp. 199-207

[45] Aiassa S, Bottino G, Mandrone G, Vigna B. Studio multidisciplinare per la valutazione della franosità di alcuni versanti collinari in Alta Langa. In: Luino F, editor. "La previsione delle catastrofi idrogeologiche: il contributo della ricerca scientifica". Proceedings of the Conference "Alba 96", Alba (Italia) 5-7 novembre 1996, vol. 2; 1998. pp. 185-198

[46] Bandis SC, Del Monaco G, Dutto D, Margottini C, Mortara G, Serafini S, Trocciola A. In: Landslides and Precipitation: the Event of 4-6th November 1994 in the Piemonte Region, North Italy. Ch. 20; 1999. pp. 315-326

[47] Mason PJ, Rosenbaum MS. Geohazard mapping for predicting landslides: An example from the Langhe Hills in Piemonte, NW Italy. Quarterly Journal of Engineering Geology and Hydrogeology. 2002;**35**(4):317-326. DOI: 10.1144/1470-9236/00047

[48] Torta D, Bottino G, Mandrone G. Valutazione delle condizioni di rischio dei centri storici delle Langhe tramite l'analisi dell'influenza delle precipitazioni sui meccanismi d'innesco dei dissesti (Cuneo-Italia); 2000. Available from: http://hdl.handle.net/11583/1408488

[49] Mandrone G, Torta D. Modello previsionale per l'innesco di frane da scivolamento planare nelle langhe: monitoraggio del livello della falda e sua correlazione con i dati meteorologici. Int. Conf. "Il territorio fragile", X Congr. Naz. Geol., Roma, 7-10 dicembre; 2000. pp. 145-154

[50] Mandrone G. Il ruolo dell'acqua nell'innesco di frane planari negli ammassi rocciosi eterogeneo delle Langhe (Italia nord-occidentale). GEAM - Associazione Georisorse e Ambiente. 2004;**XLI**(3):41-50

[51] Colombo A, Lanteri L, Ramasco M, Troisi C. Systematic GIS-based landslide inventory as the first step for effective landslide-hazard management. Landslides. 2005;**2**:291. DOI: 10.1007/s10346-005-0025-9

[52] Luino F. Sequence of instability processes triggered by heavy rainfall in northwestern Italy. Geomorphology. 2005;**66**:13-39

[53] Tiranti D, Rabuffetti D, Salandin A, Tarabra M. Development of a new translational and rotational slides prediction model in Langhe hills (north-western Italy) and its application to the 2011 March landslide event. Landslides. 2013;**10**:121. DOI: 10.1007/s10346-012-0319-7

[54] Notti D, Meisina C, Colombo A, Lanteri L, Zucca F. Studying and monitoring large landslides with persistent scatterer data. Italian Journal of Engineering Geology and Environment (IJEGE). 2013. DOI: 10.4408/IJEGE.2013-06.B-33

[55] Lollino G, Lollino P. Studio idrogeologico ed. analisi numerica di una frana di scivolamento planare. GEAM - Associazione Georisorse e Ambiente. 2001;**XXXVIII**(1):43-52

[56] ARPA Piemonte: BDGeo - Banca Dati Geologica Arpa Piemonte

Chapter 4

Characterization of Susceptible Landslide Zones by an Accumulated Index

Rocío N. Ramos-Bernal, René Vázquez-Jiménez, Sulpicio Sánchez Tizapa and Roberto Arroyo Matus

Abstract

In order to characterize the landslide susceptibility in the central zone of Guerrero State in Mexico, a spatial model has been designed and implemented, which automatically generates cartography. Conditioning factors as geomorphological, geological, and anthropic variables were considered, and as a detonating factor, the effect of the accumulated rain. The use of an inventory map of landslides that occurred in the past (IL) was also necessary, which was produced by an unsupervised detection method. Before the design of the model, an analysis of the contribution of each factor, related to the landslide inventory map, was performed by the Jackknife test. The designed model consists of a susceptibility index (SI) calculated pixel by pixel by the accumulation of the individual contribution of each factor, and the final index allows the susceptibility cartography to slide in the study area. The evaluation of the obtained map was performed by applying an analysis of the frequency ratio (FR) graphic, and an analysis of the receiver operating characteristic (ROC) curve was developed. Studies like this can help different safeguarding institutions, locating the areas where there is a greater vulnerability according to the considered factors, and integrating disaster attention management or prevention plans.

Keywords: landslide susceptibility, accumulated susceptibility index, explanatory factors, jackknife test, frequency ratio, receiver operating characteristic curve

1. Introduction

The landslides are phenomena of great importance due to their spatial dimension effects on socio-ecological systems. Typically, the landslides are present in areas where hydrometeorological phenomena also occur [1–3]. In recent years, an increase in landslide risk has been recorded due to the increase of population settlements in vulnerable areas such as coastal and mountain regions.

This problematic and global situation has aroused the interest of governments and academics, and significant efforts have been made to characterize and identify the areas that have the potential to suffer landslides. Under a territorial approach, studies and susceptibility models have been developed, where factors related to instability processes and their spatial distribution are analyzed and linked to the landslide inventories.

The remote sensing techniques and the management of geospatial information through geographic information systems (GIS) currently play a leading role in numerous studies in which statistical or semi-statistical methods are applied to the modeling and generation of landslide susceptibility cartography [4].

In the last four decades, different probabilistic methods have been developed to predict, model, assess, and produce cartography about the risk related to landslides. Van Westen et al. [5] propose three scale levels in risk maps:

A. The qualitative-heuristic approach, with empirical recognition and small-scale maps (1: 100,000–1: 250,000)

B. The statistical approach, to determine the causal factors in the quantitative susceptibility mapping (scale 1: 25,000–1: 50,000)

C. The deterministic approach, for detailed large-scale studies (1: 2000–1: 10,000)

Guzzetti et al. [6] describe that the most important proposed methods can be grouped mainly in the following categories:

1. Geomorphological risks mapping

2. Analysis of landslide inventories

3. Heuristic or indexing methods

4. Functional and statistical models

5. Geotechnical models or based on the physical structure

The literature about the evaluation of susceptibility to landslide events is vast; Carrara [7], Guzzetti [8], Chung et al. [9], Baeza and Corominas [10], Hervás et al. [11], Ayalew and Yamagishi [12], Lee [13], Mancini et al. [14], Kayastha et al. [15], and Dou et al. [16], among others, use the statistical methods of multivariate analysis (discriminant analysis, processes of analytical hierarchies or logistic regression). According to Feizizadeh and Blaschke [17], the use of these methods has increased in landslide susceptibility mapping; however, the results are sensitive to the quality of the dependent variable, for instance, the inventory of previous landslide events [4, 18, 19].

Other authors have analyzed damaged scenarios with a geostatistical approach applying kernel density maps to evaluate geological, geomorphological, and risk data [20]. Lazzari and Danese proposed a multi-temporal kernel density estimation approach for local landslide susceptibility evaluation and forecasting, based on spatial statistics techniques and in particular on kernel density estimation, considering the interaction between landslides that are located close to each other, as a second-order effect in landslide spatial distribution [21].

Due to the complexity of landslides, many authors have focused on nondeterministic methods to assess the susceptibility. Lee et al. [22], Gómez and Kavzoglu [23], Park et al. [24], and Conforti et al. [25] have applied data mining methods using artificial neural networks, while Ercanoglu and Gokceoglu [26] and Pistocchi et al. [27] have applied the fuzzy logic methods.

Ward et al. [28], Terlien et al. [29], Fall et al. [30], and Jia et al. [31] have adopted a mechanical approach, using deterministic and/or numerical methods, which have been successful for the analysis of the stability of a slope [32].

In this study, we apply a heuristic method, which considers the relationship between the location of landslide events previously occurred and the geomorphological and geological environment [33–36]. In the heuristic methods, the experts' criteria are essential, due and based on their expertise, and they determine the factors that influence the instability [37]; here each factor is weighted according to its importance/influence on the landslide triggering. A final map is obtained by superimposing the thematic layers related to each factor. Examples of the application of this type of indirect methods are described below.

Irigaray [38, 39] proposes the critical slope method, with which the susceptibility of lithology in slope intervals is estimated. The critical slope per lithological unit is obtained from the relation of the increase of the percentage occupied by the surface of rupture of the class of immediate lower slope and the increase of slope of both classes.

Van Westen [40, 41] developed a procedure that calculates the landslide density, assigning a weight to the variables (factors), through the difference with the total density and finally adding the weighted densities. An important feature is that this method contemplates different types of landslides and activity stages, considering the slid zones.

Sarkar et al. [42] performed a similar procedure, but without assigning weights, they calculate the landslide density for each class and factor, obtaining a susceptibility index through the weighted accumulation of the individual indices corresponding to each factor included and thus obtaining the final susceptibility map.

Barredo et al. [43] apply two knowledge-based methods. The direct method performs a very detailed geomorphological cartography using specifically coded polygons that were evaluated one by one by experts to assess the type and risk degree. The indirect method is based on an indexing technique, where the conditioning factors are combined using multicriteria evaluation techniques.

Hervás and Barredo [44] and Hervás et al. [11] propose an indirect method of indexing to evaluate the landslide susceptibility, through the pondered linear sum of the weights of the considered factors and classes, for the assignation of the final weight corresponding to each conditioning factor. In their study, they applied the analytical hierarchy method, creating a matrix with the relative value judgments between pairs of conditioning factors of instability. The resulting susceptibility index was presented in a five-category thematic map.

Ayalew et al. [34] applied the linear weight combination method to determine classes of control parameters (factors): lithology, slope, terrain orientation, elevation, profile, and flat curvature, considering the order of importance in the process of landslides and weighing the impact of one parameter against another.

Wati et al. [35] applied a heuristic approach with the weighted scoring method, where the judgment of an expert determines the weighting and scoring, which represents the effect of each factor in the process of a landslide. The higher the weight and the score, the higher will be the influence of a particular factor to trigger a landslide. The weight designation to the corresponding factors was aided by the Integrated Land and Water Information System (ILWIS) software, and they used a rank of values, 1 to 4. The results were presented in a thematic susceptibility map organized in five classes whose ranges were determined by the Natural Breaks method [45] based on the accumulated weight score.

The work presented here is only one phase of a global susceptibility analysis project in the Guerrero State, which includes landslides, floods, earthquake damage, and environmental effects due to industrial activity, among other phenomena that occur in the study area.

Under this background, the present work consists of the analysis, design, and implementation of a spatial model to characterize the landslide susceptible zones in the central area of Guerrero State in Mexico.

2. Study area

The study area covers a space of 3300 km^2, in the central zone of Guerrero State in Mexico, and consists of a mountainous region, with elevations ranging from 280 m to 3540 m above mean sea level and slopes over 40 (**Figure 1**).

According to the Mexican Institute of Statistics and Geography (INEGI), for the year 2013, temperature variations were recorded from 14.3°C (December) to 28.3°C (May) [46].

The records of precipitation of the National Meteorological Service (SMN-CONAGUA) for the year 2013 indicate that the ranges vary from 800 mm as minimum average to 2100 mm as maximum recorded from June to September [47]. The area was 74.8% covered by forest (coniferous, mesophilic, and mixed), 14.1% by deciduous forest, 7.8% by agricultural land, 3.2% by induced vegetation, and 0.1% by human settlements and urban areas [48]. According to the 2010 Population and Housing Census, there are 187 cities and towns, in which there are 15,230 homes inhabited by 59,098 people [49].

Geologically, the area is located physiographically in the Sierra Madre del Sur [50] and has a variety of metamorphic rock composition consisting of schists and gneisses of biotite and quartzite, outcrops of deposited limestone, metavolcanic rocks with sedimentary influence, siltstones, sandstones, conglomerates, and carbonate rocks. Rhyolitic rocks are also found as a result of Oligocene-Miocene volcanism. The youngest rocks correspond to alluvial deposits present in the margins and riverbeds [51, 52]. **Figure 2** shows the geological map of the study area, integrated from two geological charts and ten geological-mining charts, referenced to 2012.

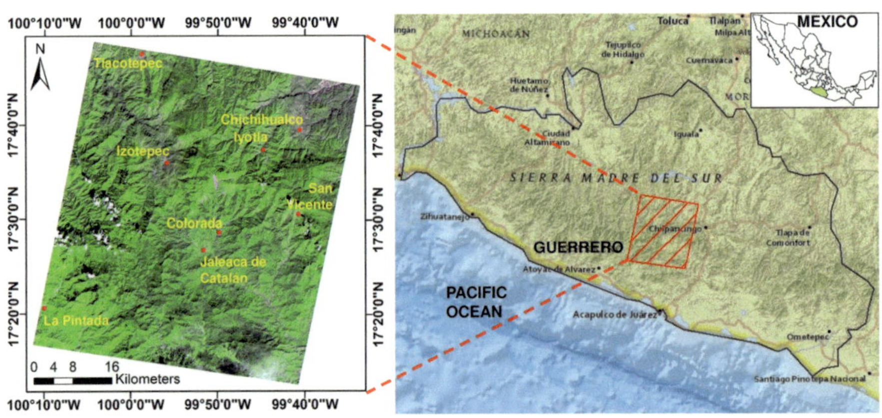

Figure 1.
Study area, the south-central region of the state of Guerrero, Mexico. Source: Adaptation of ASTER image from Dec 13 2013.

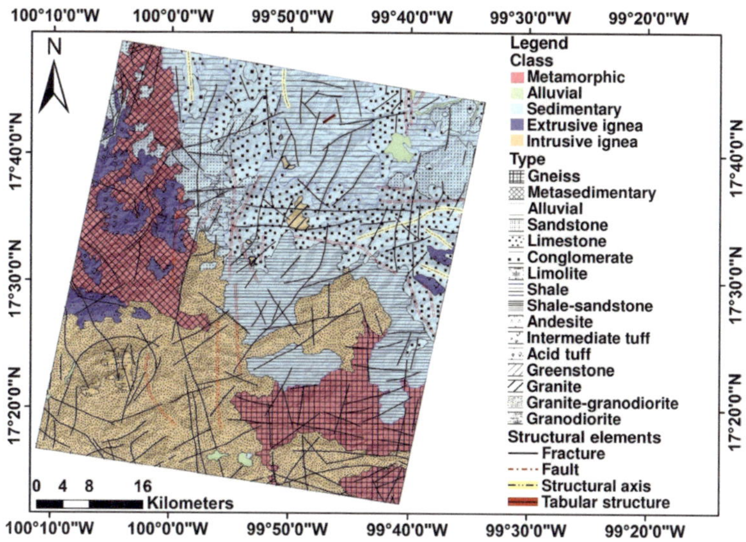

Figure 2.
Geological map of the study area [53].

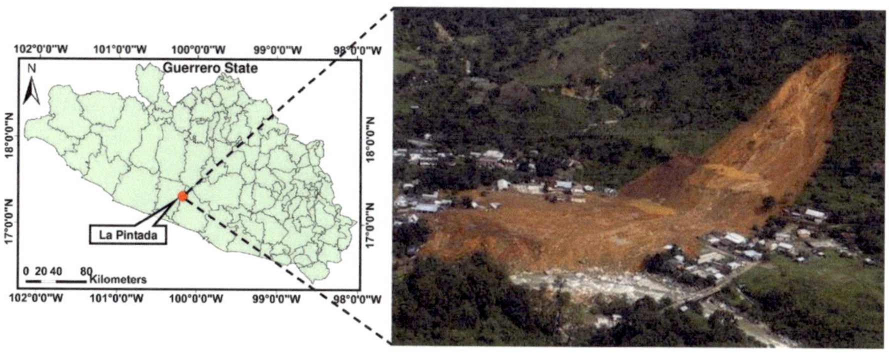

Figure 3.
A massive landslide occurred on La Pintada community. September 2013, in the study area. Source: Personal adaptation from https://www.jornada.com.mx/2013/09/24/ciencias/a03a1cie.

The study area is interesting because of its geographic location, topographic, and geological conditions and also due to the presence of extraordinary hydrometeorological phenomena in recent years, which have triggered massive landslides that have severely affected the population and infrastructures of their communities. The events of September 2013 were particularly notable during tropical depression No. 13 in the Pacific Ocean and subsequent simultaneous hurricanes: Manuel in the Pacific and Ingrid in the Gulf of Mexico caused significant floods and landslides on the coast of the Guerrero State [54] (**Figure 3**).

In the study area, there is an important concentration of inhabitants. From the 187 localities distributed in the area, 182 are considered as rural (with less than 2500 inhabitants), and their environmental conditions, the materials of their homes, and characteristics are similar to La Pintada.

3. Dataset and methodology

An overview of the process to characterize, through an accumulated index, the areas susceptible to landslides is shown in **Figure 4**.

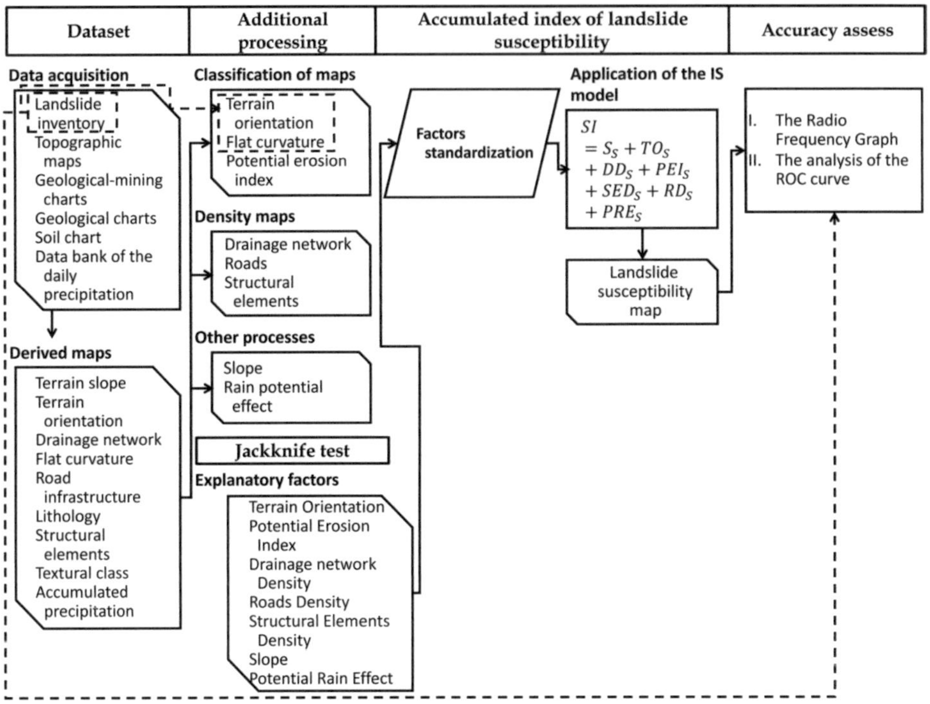

Figure 4.
Methodology overview.

As can be seen, the proposal is made up of five stages:

- Dataset. Consists of the acquisition of primary data and the generation of maps derived from it.
- Pre-processing. In this stage, additional processes are applied to the derived information, such as the categorization of maps with continuous data, the generation of kernel density maps, or the combination of maps with related information.
- Jackknife test. In this stage, the factors considered are analyzed to determine the contribution of each of them to the global susceptibility model.
- Accumulated index of landslide susceptibility. Here, the proposed model is implemented to build the landslide susceptibility map.
- Accuracy assessment. At this stage, the map generated is evaluated.

Each of the stages is explained in detail throughout the work.

3.1 Landslide inventory

An image dated on August 12, 2014, from Google Earth online platform was used, just the closest one available after the meteorological events of September 2013 previously described. Photointerpretation techniques were applied to identify and digitize a sample of 617 polygons of which 602 (17,385 pixels) represent real landslide areas and 15 (17,632 pixels) represent non-landslide zones to integrate an inventory validated through field visits in the study area (**Figure 5A**).

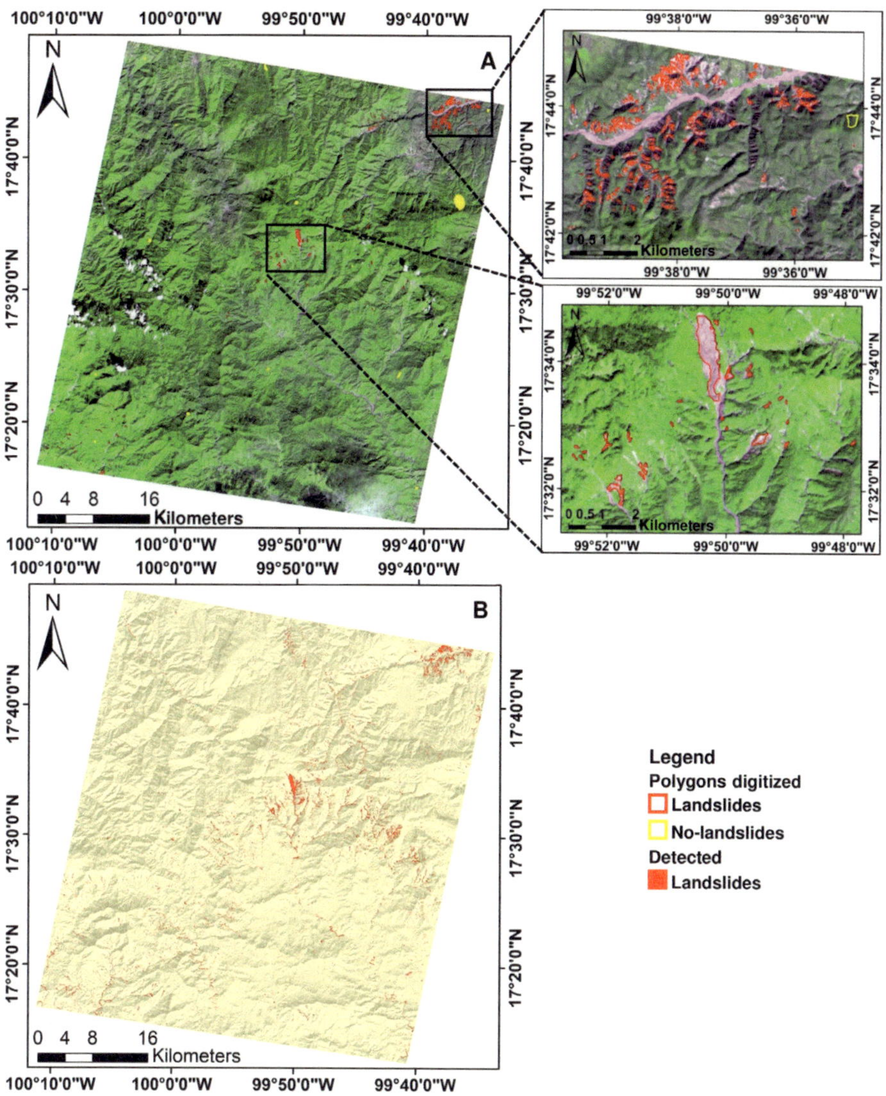

Figure 5.
Landslide inventory. (A) Landslide polygons digitized in Google earth validated. (B) landslide inventory map resulting from unsupervised detection methods [53].

Due to the conditions of access and insecurity prevailing in the study area, not all the digitized polygons were field verified. The field verification was in 321 polygons (53.3%) in secure zones near to the largest cities and close to the most important roads. The field verification indicated that 320 were translational landslides, and only 1 was a debris avalanche.

The size of the slid polygons in the integration of the landslide inventory was 450 m^2, which is the area that corresponds to two pixels, due to the spatial resolution of the ASTER images used, which is 15 m.

A landslide inventory map (**Figure 5B**) was used to assess the resulting susceptibility landslide model. This inventory was automatically produced through unsupervised detection methods applied to ASTER-derived products [53].

The described landslide inventory produced by unsupervised detection methods consists of 216,174 pixels that represent the landslide zones caused by the extraordinary rains of the simultaneous hurricanes Ingrid and Manuel in September 2013.

These pixels were divided into three equal parts, of which 2/3 (144,116 pixels) were used to define the degree of contribution of the factor terrain orientation and flat curvature to the susceptibility model.

The other part (72,058 pixels) was complemented by a similar number of randomly selected pixels of the inventory map that were identified as non-landslides and reserved for the stages of evaluation of the influence of factors on the occurrence of landslides and to assess the accuracy of the landslide susceptibility map obtained in this work.

3.2 Conditioning and detonating factors

In this work, the quantity and type of the analyzed factors were the results of a selection process supported on the expertise or researchers based on specific characteristics of each factor and the conditions of the study area. Thus, it is important to emphasize that finally, the explanatory factors are considered, depending on the characteristics of the study area and the availability of corresponding information. The factors considered, as well as the processes applied for their integration, are described as follows.

Based on the data obtained from the topographic maps, courtesy of National Institute of Statistics and Geography of Mexico (INEGI), a 15 m grid cell digital elevation model (DEM) was generated; then, maps of the angular slope, terrain orientation, drainage network, and flat curvature were produced from DEM to be used as factors in the susceptibility modeling.

From the topographic maps, the road infrastructure data was obtained. The lithology map was obtained from the geological-mining charts, while the map of the structural elements was obtained from the geological chart courtesy of INEGI and the Mexican geological service (SGM).

From the soil chart of Guerrero State (INEGI), the textural class map of the study area was obtained. Also, a map of accumulated precipitation was generated from the data bank of the daily precipitation from January to September 2013, obtained from the meteorological stations located in or near to the study area (**Figure 6**). With the data of accumulated precipitation, textural class, and terrain slopes, a new combined product was generated that was called the potential rain effect map.

According to the kind of data of each factor, in some cases, additional processing was performed to be used in the modeling of susceptibility. For the terrain slopes (S), in concordance with Alcantara-Ayala [55], the slope values smaller than 5° were considered as flat zones, and then without effect on the susceptibility index, while for the upper slope values, a limit value through the statistical parameters of all slope data was determined, calculated by the mean plus twice the standard deviation ($\mu + 2\sigma$). The resulting value was 45.3°, and from this value onwards, it was considered that the effect on the susceptibility to slip is the same, and thus all the upper-value slopes were equalized considering them with the same weight (**Figure 7A**) [56].

About lithology, a susceptibility degree was defined for each lithological unit. The map was categorized according to Aramburu Maqua and Escribano Bombín [57], who refers to the degree of cohesion, potential erosion, and mechanical behavior of each type of rock.

In this sense, the potential erosion index (PEI) was used, since it refers to erosion by surface runoff due to hydrological factors, which are considered in this study as the primary trigger of landslides. Based on the analysis of the lithological

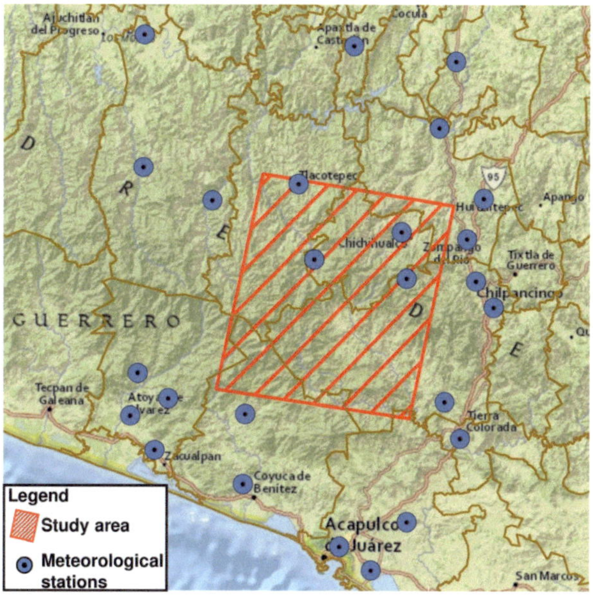

Figure 6.
Meteorological stations considered in the study. Source: Personal adaptation from ESRI and SMN-CONAGUA.

content of the rocks present in the area, five classes were defined that indicate the susceptibility degree, related to the potential erosion of the rock type. According to the categorization based on the PEI [57] and the identified rock types, four susceptibility classes were determined: (1) *very low*, (2) *low*, (3) *medium*, and (5) *very high* (**Figure 7B**). It is important to note that the classification corresponds to the cited works consulted, and according to the analysis performed, class (4) *high* does not exist in the study area.

Regarding the drainage network maps, road infrastructure, and geological structures, it was considered to use density maps, since if a point in the space next to an "element" (runoff, road, or geological structure) has a certain degree of susceptibility, then this should be greater if the same point is close to two or more elements; therefore, a density map would represent the most susceptible areas to slide according to its position with respect to one or more of the elements considered.

From each map (drainage network, road infrastructure, and geological structures), the density maps were generated by the kernel density estimator (KDE), which calculates a magnitude of influence of the "elements" per unit area according to the neighborhood of the pixels [20, 58]. The KDE can be applied on specific entities or lines. KDE implies the use of parameters, among which the search radius (bandwidth) is one of the most important and represents the value of the radius within which the density will be estimated.

The default search radius (SR) is computed according to the spatial distribution of input dataset to be analyzed using a spatial variant of Silverman's rule of thumb that is robust to spatial outliers (i.e., points that are far away from the rest of the points), through the equation

$$SR = 0.9 * min\left(SD\sqrt{\frac{1}{\ln(2)} * Dm}\right) * n^{-2} \quad (1)$$

Conceptually, an adjustment of a uniform curved surface is made on each "element." The value of KDE is higher on the line and decreases as the pixels are farther

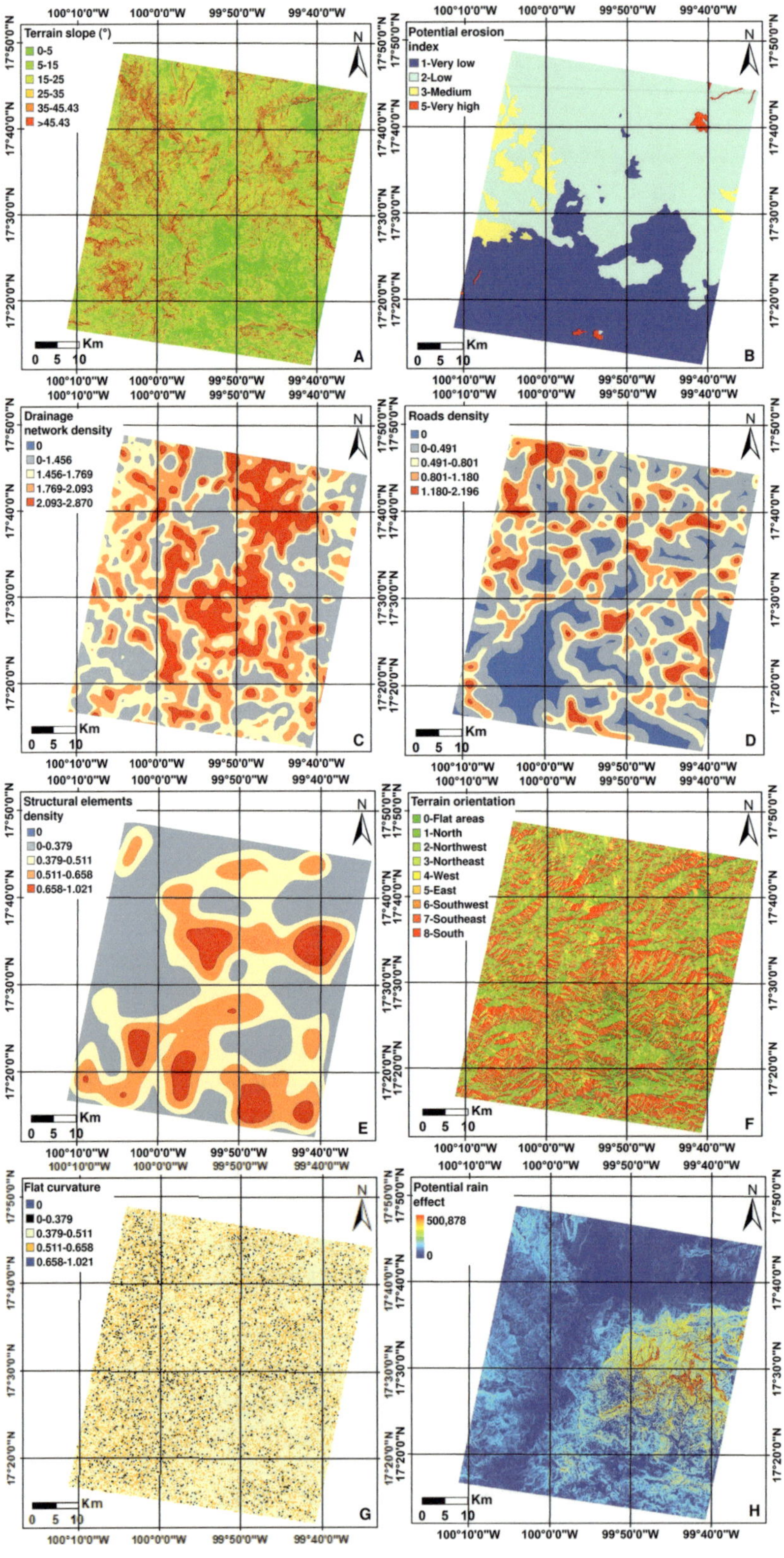

Figure 7.
Explanatory factors of landslide susceptibility in the central zone of Guerrero state. (A) Terrain slope. (B) Potential erosion index. (C) Drainage network density. (D) Road density. (E) Structural element density. (F) Terrain orientation. (G) Flat curvature. (H) Potential rain effect.

from the feature, even reaching the zero value. To calculate the density of each pixel, the values of all the superposed kernel surfaces are added [59] (**Figure 7C–E**).

The terrain orientation (TO) was classified according to Rawat et al. [60], excluding the value of −1 since it represents flat areas and does not contribute any degree of susceptibility to landslides in the model. The rest of the orientation values were organized in eight classes, to which a cross-tabulation was applied with 2/3 of the slides taken from the ground truth (landslide inventory map) to determine the frequency of each class of orientation of the terrain in which the pixels are slid.

From the crossing of information of the class of terrain orientation and the pixels of the landslide inventory map, the map of orientations was reclassified according to the frequency of the slid pixels in each class, being the order of frequency and the degree of susceptibility assigned (**Table 1**, **Figure 7F**).

The flat curvature (FC) is an important topographic feature commonly used in landslide susceptibility studies [61]. The slopes can be subdivided into regions of flat curvature, lateral concave, lateral convex, and linear; however, statistical analysis of the flat curvature and landslide datasets indicates that slopes with linear flat curvature have the highest probability of landslides, and this probability decreases as the slopes become more concave or convex [62].

The flat curvature map contains values from −25,798 to 30,944, which were adjusted according to the empirical information of the landslide inventory map, using a similar methodology to that applied in the terrain orientation; first, the curvature values were analyzed, and then the data were cross analyzed with the landslide inventory information, using two-third of the slid pixels considered as ground truth. Thus, the sampled slid pixels show curvature values from −12,427 to 5486; to these range values, a classification was applied in three ranges by the Natural Breaks method [45]; and for each class, the frequency of slid pixels was obtained (**Table 2**).

Most of the sampled pixels (72.1%) are included in the range (−1.328 to 0.358) that corresponds to class 2; therefore this class was assigned with the highest degree of susceptibility, and according to the frequency values, to class 1 the *medium* susceptibility was assigned, and to class 3 the *low* susceptibility was assigned. For curvature values outside of the analyzed ranges [values lower than −12,427 (convex areas) and higher than 5486 (concave zones)], a null susceptibility was assigned for effects of the development of the susceptibility model; this is because they were not linked to slid pixels, according to the empirical analysis (**Figure 7G**).

N.	Class	Azimuth terrain orientation (°)	Frequency (%)	Reclassification
1	North	337.5–22.5	2.48	1
2	Northeast	22.5–67.5	5.42	3
3	East	67.5–112.5	13.95	5
4	Southeast	112.5–157.5	19.74	7
5	South	157.5–202.5	25.36	8
6	Southwest	202.5–247.5	17.68	6
7	West	247.5–292.5	11.31	4
8	Northwest	292.5–337.5	3.60	2

Source: [56].

Table 1.
Reclassification of terrain orientation according to slid pixel frequency.

Class	Minimum	Maximum	Pixels %
1	−12.427	−1.328	10.8
2	−1.328	0.358	72.1
3	0.358	5.486	17.0

Source: [56].

Table 2.
Flat curvature and relative frequency from the sampled slide pixels.

The potential rain effect (PRE) map (**Figure 7H**) did not require any empirical analysis.

3.2.1 Evaluation of the influence of factors on the occurrence of landslides: the jackknife test

Once the information of the considered factors in the study was integrated, the Jackknife test was applied, which allows identifying the importance degree of each factor by executing the models in a separate way, with different sets of factors and excluding the factor analyzed. The Jackknife test provides alternative estimators of landslide patterns as a function of the factors, allowing so the identification of the significance of each of them, by their contribution degree and by their importance regarding the global results of the model [63–66].

In this study, the Jackknife test graph was produced as a useful tool to assess the individual contribution of each factor used and also to assess the contribution of each range or category of each factor to the susceptibility landslide.

3.3 Accumulated index of landslide susceptibility

The applied model consists of the integration of an accumulated global index of landslide susceptibility, calculated from the individual contributions of each factor in the following described stages.

3.3.1 Factors standardization

Each factor reports continuous or discrete values. Therefore, a standardization process was applied; in this way, individual contribution values were assigned to the susceptibility model, calculated between 0 (null contribution) and 1 (maximum contribution); and thus, evaluate the corresponding contribution in the global accumulated susceptibility index. So, all the factors will have the same weight in the model.

The standardization process was made according to the data of each factor. Thus, for instance, for the PEI factor which is represented in a thematic map of five categories of contribution to susceptibility ((1) *very low*, (2) *low*, (3) *medium*, (4) *high*, and (5) *very high*), the standardized value was determined by finding the relationship of each pixel in connection with the value corresponding to the highest category, resulting in the PEI factor with values between 0 and 1.

For the standardization of some other factors, such as the flat curvature or the terrain orientation, the landslide inventory map was used to empirically obtain range values of curvature or orientation in slid areas (ground truth). A schematic example of the terrain orientation standardization process is shown in (**Figure 8**).

The standardization of the factors registered in maps of continuous values was determined by finding the relationship between the value of each pixel and the highest pixel value recorded.

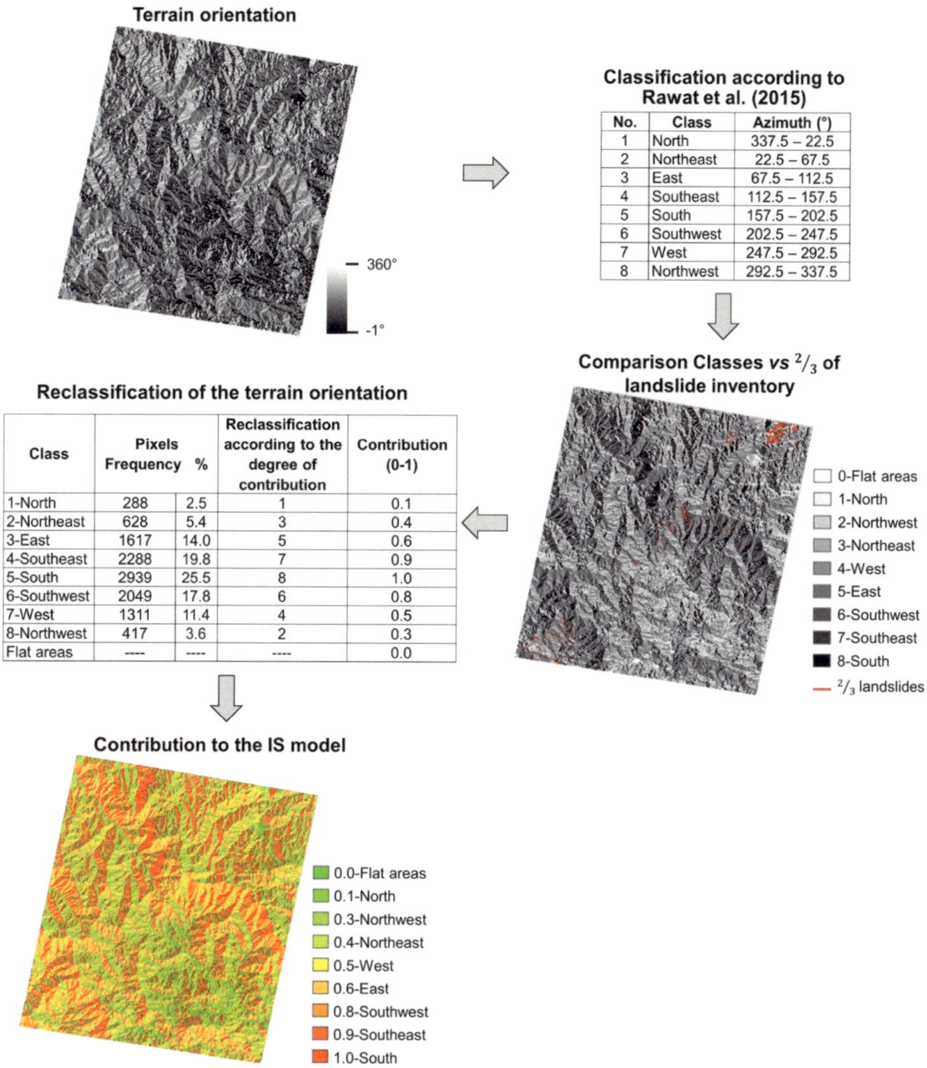

Figure 8.
Terrain orientation standardization process.

All the models and processes described in this paper were developed using the Dinamica EGO software [67–69].

3.3.2 Landslide susceptibility map through the susceptibility index (SI)

The final value of the SI is integrated by the accumulation of the individual contribution of each factor considered in the study [70], calculated by the equation:

$$SI = S_S + TO_S + DD_S + FC_S + PEI_S + SED_S + RD_S + PRE_S \quad (2)$$

where
SI = Cumulated landslide susceptibility index.
S_S = Standardized terrain slope.
TO = Standardized terrain orientation.
DD_s = Standardized drainage network density.

FC_S = Standardized flat curvature.
PEI_S = Standardized potential erosion index.
SED_S = Standardized geological structural elements density.
RD_S = Standardized roads density.
PRE_S = Standardized potential rain effect.

Under the premise that each factor provides a degree of susceptibility with standardized values ranging from 0 (null) to 1 (high), the final map obtained represents the pixel by pixel accumulated index of landslide susceptibility of the analyzed area, according to the considered factors. It is important to note that from the obtained results of the Jackknife test, the accumulated SI model can be modified by excluding some factors due to its degree of contribution.

3.4 Accurate assessment of the accumulated index of landslide susceptibility

The accuracy in the susceptibility mapping refers to the degree of agreement between the prediction of areas susceptible to landslides obtained by the applied methods and the location of actual landslides, according to the information considered as reference data or ground truth in the landslides inventory.

The implementation of the accumulated SI model produces a susceptibility map with continuous values, in which, according to their ranks, applying natural breaks were categorized into five classes: (1) *very low*, (2) *low*, (3) *medium*, (4) *high*, and (5) *very high*. On the other hand, there is a landslide inventory map organized into two categories: (1) *non-landslides* and (2) *landslide*. So, to determine the accumulated SI model accuracy, two methods were used:

I. The radio frequency graph (RFG). The frequency ratio was graphed, taking the slid pixels from the inventory sample reserved for the validation stage and superimposing them on the landslide susceptibility map resulting from the applied SI method. We identify and quantify the coincidence frequency of slide pixels in both: the map produced by the model and landslide inventory map, for each class considered in the model (*very low*, *low*, *medium*, *high*, and *very high*) [71]. In an ideal landslide susceptibility map, the frequency value should be increased from a low susceptibility to a very high susceptibility [72, 73].

II. The analysis of the receiver operating characteristic (ROC) curve. It is a standard method to evaluate the accuracy of a diagnostic test [74]. The ROC curve is a comparative graphical representation of ordered pairs between the false-negative and false-positive rates, resulting in a diagnosis for classes of values. Conventionally, the graph shows the rate of the pixels diagnosed as false-positive (false-positive rate (FPR)) on the X-axis (Eq. (3)) and the rate of the pixels diagnosed as true positive (true-positive rate (TPR)) on the Y-axis (Eq. (4)):

$$X = FRP = 1 - \left[\frac{TN}{TN + FP}\right] \tag{3}$$

$$Y = TRP = 1 - \left[\frac{TP}{TP + FN}\right] \tag{4}$$

where
TN represents the true-negative pixels.
FP represents false-positive pixels.

TP represents the true-positive pixels.

FN represents false-negative pixels.

The area under the ROC curve (AUC) characterizes the quality of the predicted system, describing its ability to anticipate the occurrence or nonoccurrence of predefined events. The value of the AUC varies from 0.5 to 1.0. If a model does not predict the occurrence of landslides better than randomly, then the AUC would be equal to a value of 0.5, while a ROC curve with an AUC value of 1 represents the perfect prediction. The quantitative-qualitative relationship between the AUC and the prediction accuracy can be classified as follows: values from 0.9 to 1, excellent; from 0.8 to 0.9, very good; from 0.7 to 0.8, good; from 0.6 to 0.7, medium; and finally from 0.5 to 0.6, poor [75].

In the application of accuracy assessment, one-third of the total of pixels identified as a landslide in the inventory map (72,058 pixels) were used, which were omitted from the previous stages of training and preparation of the factors. This information was supplemented with an equal number of pixels randomly selected from the landslide inventory map that were identified as non-landslides.

4. Results and discussion

4.1 Analysis of the jackknife test

As we explained before, the Jackknife test allows identifying the degree of importance of each considered factor, executing the model in a separate way with different sets of factors and excluding each factor in turn. **Figure 9** shows the results of the Jackknife test for the factors used in the present study.

The graph shows in blue the individual contribution of each factor in the implementation of the model; in cyan color, the gain of the model is shown, excluding the factor analyzed; and finally, the bar in red represents the gain of the model including all the factors.

According to the graph, we can see that the factor with the highest individual contribution corresponds to the potential erosion index, followed by terrain orientation and terrain slope factors. Also, we can see that the factor with the lowest contribution to the model corresponds to the flat curvature and can also be verified that the gain of the model is not affected if the flat curvature is omitted in the execution of the model.

One of the most important contributions of the Jackknife test is the possibility of identifying the degree of contribution and importance of each of the analyzed factors. Thus, we can determine which of them are essential in the reproduction of the pattern represented by the random sample of the landslides observed in the ground truth and so determine its degree of importance concerning the results reported by the global model. Thus, it is also possible to identify those factors that do not substantively contribute individual gains to the global model and observe the gain that is achieved when they are excluded. It is also possible to consider eliminating them in the model implementation, knowing previously the result that would be obtained when doing so. So, it is theoretically possible that the accuracy of the results achieved by the global model remains the same, and it is even likely that there will be improvements in the overall result by excluding them from the model in a new performance.

According to the results of the Jackknife test (**Figure 9**), it may be possible to propose the elimination of the flat curvature in the application of the model, since it is the factor that records the lowest individual contribution to the global result and at the same time is the factor that if omitted in the execution would not affect the overall result of the model.

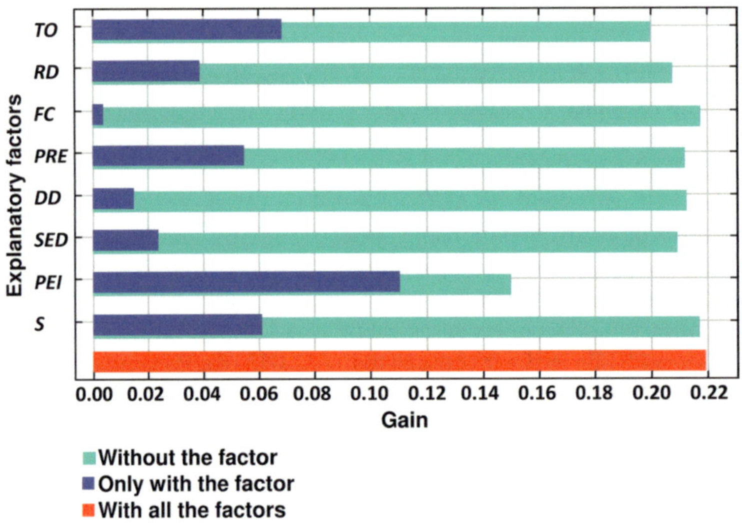

Figure 9.
Jackknife test result graph.

4.2 Accumulated SI model

The implementation of the cumulative global model resulted in the landslide susceptibility map shown in **Figure 10**.

The resulting map is represented in continuous values, which according to their ranks was categorized into five classes; thus, the areas identified as (1) *very low* and (2) *low* susceptibility can be seen in blue tones; in shades of yellow are the areas with category (3) *medium* and in orange and red the areas evaluated by the model with categories (4) *high* and (5) *very high* susceptibility.

A concentration of areas with the highest susceptibility levels can be seen in the north and center of the study area, while the zones with low and very low susceptibilities are concentrated towards the south.

Analyzing the behavior of the factors in the zones identified with high susceptibility, we can note that in these areas the slopes have a terrain orientation to the south, southeast, and southwest (**Figure 7F**), which favor the occurrence of landslides in the study area. On the other hand, in those same areas identified as having high susceptibility, there are slopes greater than 30° (**Figure 7A**), which further favors, in addition to the orientation, the presence of landslides.

It should be noted that the SI method considers the logical, empirical, and even intuitive effect designated to each factor, that is, higher terrain slopes correspond to greater susceptibility; greater proximity to faults also corresponds to greater susceptibility. In other words, the weight is not assigned to the factors according to the presence of landslides but mainly is assigned as a function of the characteristics and factor attributes, although, as previously explained, in some specific cases the contribution of susceptibility index was determined empirically according to the presence of landslides.

Figure 11 shows a zoom to the northeast zone of the study area, which is characterized by having recorded a significant presence of landslides occurred in September 2013.

In the figure, an acceptable degree of accuracy is observed in the susceptibility map generated by the SI model, since in general, the occurrence of landslides coincides with the areas categorized with high or very high susceptibility.

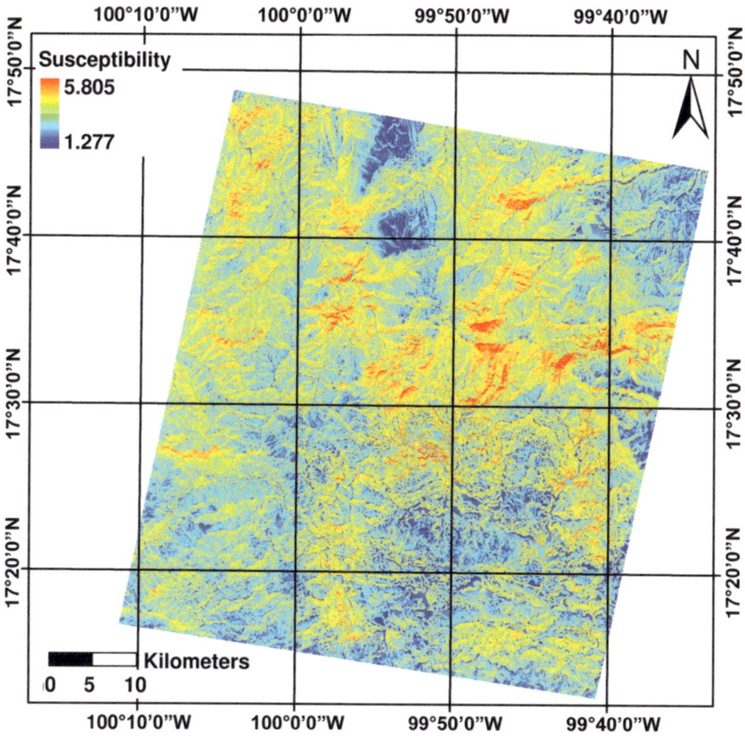

Figure 10.
Map of the cumulative susceptibility index model.

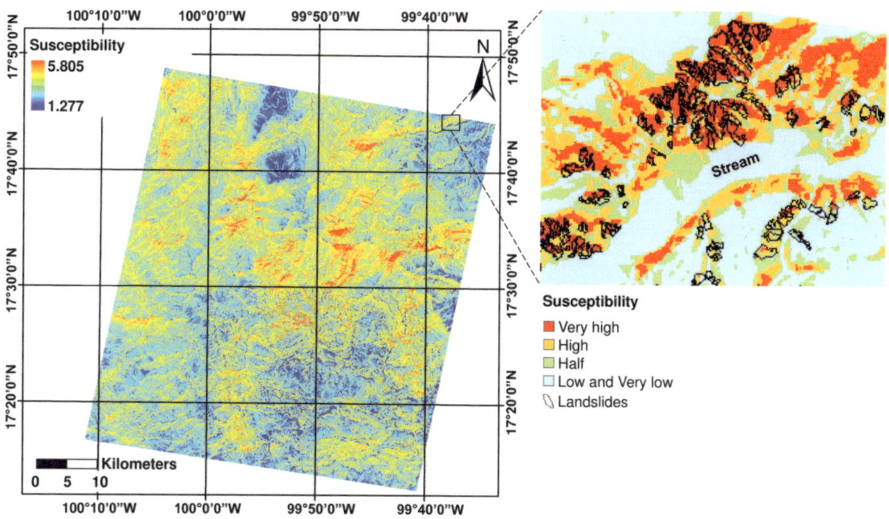

Figure 11.
Massive landslide zone occurred in 2013.

It can be observed that most of the landslides occurred to the north of the stream are over an area identified by the model in the very high susceptibility class, and the few landslides that did not meet this condition occurred in another area identified as high susceptibility. On the other hand, also it can be seen that most of the landslides located to the south of the stream occurred over areas identified as high susceptibility, but it is also possible to see landslides recorded in the ground truth that occurred in areas identified in the medium susceptibility category.

4.3 Accuracy assessment of the accumulated SI model

4.3.1 Analysis of the radio frequency graph (RFG)

Following the methodology for the accuracy assessment, the sampled pixels categorized as landslides from the map inventory reserved for validation were superimposed on the resulting susceptibility map from the implemented model, and then the frequency of the occurrence of landslides was obtained on each map category, building the graph of the frequency radius (**Figure 12**).

It can be observed that the pattern of the resulting graph shows that the frequency of the slid pixels of the ground truth is lower in the *very low* susceptibility category, and this value gradually increases over the following categories to finally reach the highest frequency value in the *very high* category; in concordance with Pradhan and Lee [72], Pourghasemi et al. [73] and Pradhan and Lee [76] in their work conclude that in an ideal map of landslides susceptibility, the value of the slid frequency should gradually increase from a low susceptibility zone to very high one.

If we group the extreme categories, the *very low* and *low* classes in one hand, and the *high* and *very high* classes on the other, we confirm that the susceptibility map shows an accumulated frequency of 15.4% of pixels slid in the *very low* and *low* susceptibility categories, and on the other end (*high* and *very high* susceptibility categories), we can see that the susceptibility map shows an accumulated frequency of 68.1% of pixels slid. This fact confirms the positive trend of the results obtained in congruence with the previously cited works.

4.3.2 Analysis of the receiver operating characteristic (ROC) curve

From the cross-validation between the categories included in the SI model, and the random sample of the ground truth taken from the landslide inventory map,

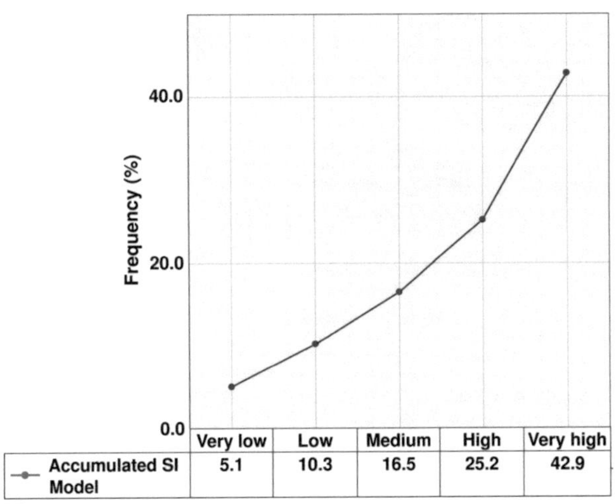

Figure 12.
Frequency ratio graph.

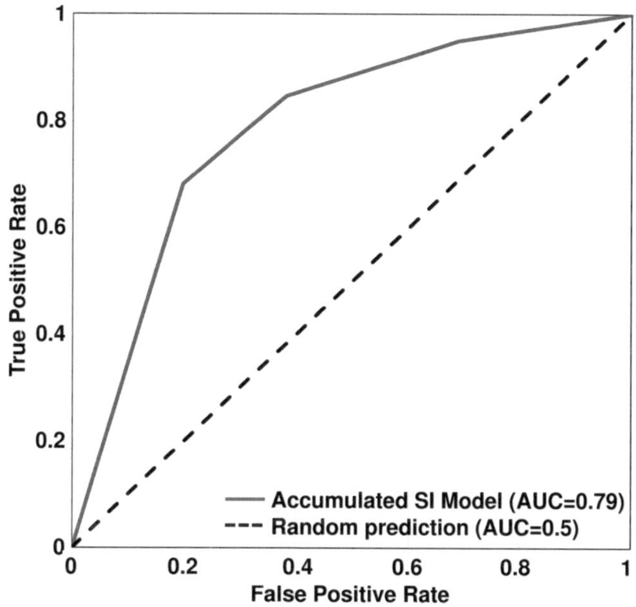

Figure 13.
ROC curve and AUC from the accumulated SI model.

including for this case both landslides and non-landslides, the graph corresponding to the ROC curve was constructed, and the area under the curve (AUC) was calculated (**Figure 13**).

As previously mentioned, the AUC describes the ability of a model to predict the occurrence or nonoccurrence of predefined events. According to Chen et al. [77], the higher AUC represents the higher performance of the model analyzed.

The AUC value reported by the model is 0.79, indicating that its predictive capacity is good, under the classification values consulted in the work of Yesilnacar [75], Devkota et al. [78], and Kim et al. [79].

According to the results, the SI accumulated model, in the way that has been implemented in the present work, has proven to be a valid method to characterize the areas susceptible to the occurrence of landslides in the study area. However, one of the main problems consists in assigning the correct weight value to each of the explanatory factors considered. According to Barredo et al. [43] and Clerici et al. [80], the heuristic methods use selective criteria that require expert knowledge to be applied properly, which implies a substantial degree of subjectivity insofar as each factor is assigned with a certain degree of a priori importance.

It is important to highlight the design of the model to be implemented and the explanatory factors to be considered depending on the specific characteristics of the study area and the availability of the corresponding data. In other words, the results shown here may be different if the described methodology and the same amount and type of explanatory factors are applied in a study area with characteristics different from those considered in the present study. However, the methodological proposal can be taken as a starting point and as a guide for the specific design of a model applied to a different study area and adjusted according to the available information and the topographic, geological, hydrological, or environmental conditions prevailing in that area.

5. Conclusions

According to the Jackknife test, lithology is the factor that records the most significant individual contribution to the global susceptibility, followed by the orientation and slope of the terrain. On the other hand, the factor with the least contribution to the model is the flat curvature.

According to the results obtained by the SI method, the heuristic approach proved to be valid and straightforward, in agreement with similar works [43], to assess the susceptibility of landslides on a medium scale such as the central zone of the State of Guerrero.

Cartographic information about landslide susceptibility is necessary for areas prone to landslides since it can be used as a support to the establishment of early warning systems for localities and residents in risk areas. It can also be a support to the institutions and organizations responsible for safeguarding in the management of events of this type, locating the areas with a higher vulnerability according to the factors considered and integrating care or disaster prevention plans considering areas with higher priority than others. It is also important to mention that this work can be complemented with the definition of evacuation routes, identification of care centers, and identification and capacity of shelters, among other possible actions, all this putting as a fundamental focus, the knowledge of the territory through the application of geographic information technologies.

Author details

Rocío N. Ramos-Bernal, René Vázquez-Jiménez*, Sulpicio Sánchez Tizapa and Roberto Arroyo Matus
Academic Group UAGro CA-93 Natural Risks and Geotechnology, Engineering Faculty, Guerrero Autonomous University, Chilpancingo, Guerrero, Mexico

*Address all correspondence to: rvazquez@uagro.mx

References

[1] Larsen MC, Torres-Sánchez AJ. The frequency and distribution of recent landslides in three montane tropical regions of Puerto Rico. Geomorphology. 1998;**24**(4):309-331. DOI: 10.1016/S0169-555X(98)00023-3

[2] Guthrie R, Evans S. Analysis of landslide frequencies and characteristics in a natural system, coastal British Columbia. Earth Surface Processes and Landforms: The Journal of the British Geomorphological Research Group. 2004;**29**(11):1321-1339. DOI: 10.1002/esp.1095

[3] Remondo J, Soto J, González-Díez A, Terán d, Ramón Díaz J, Cendrero A. Human impact on geomorphic processes and hazards in mountain areas in northern Spain. Geomorphology. 2005;**66**(1-4):69-84. DOI: 10.1016/j.geomorph.2004.09.009

[4] Poiraud A. Landslide susceptibility–certainty mapping by a multi-method approach: A case study in the tertiary basin of Puy-en-Velay (massif central, France). Geomorphology. 2014;**216**:208-224. DOI: 10.1016/j.geomorph.2014.04.001

[5] Van Westen CJ, Rengers N, Terlien M, Soeters R. Prediction of the occurrence of slope instability phenomena through GIS-based hazard zonation. Geologische Rundschau. 1997;**86**(2):404-414. DOI: 10.1007/s005310050149

[6] Guzzetti F, Carrara A, Cardinali M, Reichenbach P. Landslide hazard evaluation: A review of current techniques and their application in a multi-scale study, Central Italy. Geomorphology. 1999;**31**(1):181-216. DOI: 10.1016/S0169-555X(99)00078-1

[7] Carrara A. Multivariate models for landslide hazard evaluation. Journal of the International Association for Mathematical Geology. 1983;**15**(3):403-426. DOI: 10.1007/BF01031290

[8] Guzzetti F. Landslide hazard and risk by GIS-based multivariate models. In: International Workshop GIS in Assessing Natural Hazards. 1993

[9] Chung CF, Fabbri AG, Van Westen CJ. Multivariate regression analysis for landslide hazard zonation. Geographical Information Systems in Assessing Natural Hazards. 1995;**5**:107-133. DOI: 10.1007/978-94-015-8404-3_7

[10] Baeza C, Corominas J. Assessment of shallow landslide susceptibility by means of multivariate statistical techniques. Earth Surface Processes and Landforms. 2001;**26**(12):1251-1263. DOI: 10.1002/esp.263

[11] Hervás J, Barredo Cano JI, Lomoschitz A. Elaboración de mapas de susceptibilidad de deslizamientos mediante SIG, teledetcción y métodos de evaluación multicriterio. Aplicación a la depresión de Tirajana (Gran Canaria). In: Ayala-Carcedo FJ, Corominas J, editors. Mapas de susceptibilidad a los movimientos de ladera con técnicas SIG. Fundamentos y aplicaciones en España, Instituto Geológico y Minero de España; 2002. pp. 169-180

[12] Ayalew L, Yamagishi H. The application of GIS-based logistic regression for landslide susceptibility mapping in the Kakuda-Yahiko Mountains, Central Japan. Geomorphology. 2005;**65**(1-2):15-31. DOI: 10.1016/j.geomorph.2004.06.010

[13] Lee S. Application of logistic regression model and its validation for landslide susceptibility mapping using GIS and remote sensing data. International Journal of Remote Sensing. 2005;**26**(7):1477-1491. DOI: 10.1080/01431160412331331012

[14] Mancini F, Ceppi C, Ritrovato G. GIS and statistical analysis for landslide susceptibility mapping in the Daunia area (Italy). Natural Hazards and Earth System Sciences. 2010;**10**:1851-1864. DOI: 10.5194/nhess-10-1851-2010

[15] Kayastha P, Dhital MR, De Smedt F. Application of the analytical hierarchy process (AHP) for landslide susceptibility mapping: A case study from the Tinau watershed, West Nepal. Computational Geosciences. 2013;**52**:398-408. DOI: 10.1016/j.cageo.2012.11.003

[16] Dou J, Bui DT, Yunus AP, Jia K, Song X, Revhaug I, et al. Optimization of causative factors for landslide susceptibility evaluation using remote sensing and GIS data in parts of Niigata, Japan. PLoS One. 2015;**10**(7):e0133262. DOI: 10.1371/journal.pone.0133262

[17] Feizizadeh B, Blaschke T. An uncertainty and sensitivity analysis approach for GIS-based multicriteria landslide susceptibility mapping. International Journal of Geographical Information Science. 2014;**28**(3):610-638. DOI: 10.1080/13658816.2013.869821

[18] Carrara A, Crosta G, Frattini P. Geomorphological and historical data in assessing landslide hazard. Earth Surface Processes and Landforms. 2003;**28**(10):1125-1142. DOI: 10.1002/esp.545

[19] Galli M, Ardizzone F, Cardinali M, Guzzetti F, Reichenbach P. Comparing landslide inventory maps. Geomorphology. 2008;**94**(3):268-289. DOI: 10.1016/j.geomorph.2006.09.023

[20] Danese M, Lazzari M, Murgante B. Kernel density estimation methods for a geostatistical approach in seismic risk analysis: The case study of Potenza hilltop town (southern Italy). In: International Conference on Computational Science and Its Applications. 2008. pp. 415-429

[21] Lazzari M, Danese M. A multi temporal kernel density estimation approach for new triggered landslides forecasting and susceptibility assessment. Disaster Advances. 2012;**5**:100-108

[22] Lee S, Ryu J, Won J, Park H. Determination and application of the weights for landslide susceptibility mapping using an artificial neural network. Engineering Geology. 2004;**71**(3-4):289-302. DOI: 10.1016/S0013-7952(03)00142-X

[23] Gómez H, Kavzoglu T. Assessment of shallow landslide susceptibility using artificial neural networks in Jabonosa River basin, Venezuela. Engineering Geology. 2005;**78**(1-2):11-27. DOI: 10.1016/j.enggeo.2004.10.004

[24] Park S, Choi C, Kim B, Kim J. Landslide susceptibility mapping using frequency ratio, analytic hierarchy process, logistic regression, and artificial neural network methods at the Inje area, Korea. Environment and Earth Science. 2013;**68**(5):1443-1464. DOI: 10.1007/s12665-012-1842-5

[25] Conforti M, Pascale S, Robustelli G, Sdao F. Evaluation of prediction capability of the artificial neural networks for mapping landslide susceptibility in the Turbolo River catchment (northern Calabria, Italy). Catena. 2014;**113**:236-250. DOI: 10.1016/j.catena.2013.08.006

[26] Ercanoglu M, Gokceoglu C. Assessment of landslide susceptibility for a landslide-prone area (north of Yenice, NW Turkey) by fuzzy approach. Environmental Geology. 2002;**41**(6):720-730. DOI: 10.1007/s00254-001-0454-2

[27] Pistocchi A, Luzi L, Napolitano P. The use of predictive modeling techniques for optimal exploitation of spatial databases: A case study in landslide hazard mapping with expert

system-like methods. Environmental Geology. 2002;**41**(7):765-775. DOI: 10.1007/s002540100440

[28] Ward TJ, Li R, Simons DB. Use of a mathematical model for estimating potential landslide sites in steep forested drainage basins. IAHS Publication. 1981;**132**:21-41

[29] Terlien MTJ, Van Westen CJ, Van Asch TWJ. Deterministic modelling in GIS-based landslide hazard assessment. In: Carrara A, Guzzetti F, editors. Geographical Information Systems in Assessing Natural Hazards. 1st ed. Dordrecht: Springer; 1995. pp. 57-77. DOI: 10.1007/978-94-015-8404-3_4

[30] Fall M, Azzam R, Noubactep C. A multi-method approach to study the stability of natural slopes and landslide susceptibility mapping. Engineering Geology. 2006;**82**(4):241-263. DOI: 10.1016/j.enggeo.2005.11.007

[31] Jia N, Mitani Y, Xie M, Djamaluddin I. Shallow landslide hazard assessment using a three-dimensional deterministic model in a mountainous area. Computers and Geotechnics. 2012;**45**:1-10. DOI: 10.1016/j.compgeo.2012.04.007

[32] Wu Y, Chen L, Cheng C, Yin K, Török Á. GIS-based landslide hazard predicting system and its real-time test during a typhoon, Zhejiang Province, Southeast China. Engineering Geology. 2014;**175**:9-21. DOI: 10.1016/j.enggeo.2014.03.005

[33] Anbalagan R, Singh B. Landslide hazard and risk assessment mapping of mountainous terrains—A case study from Kumaun Himalaya, India. Engineering Geology. 1996;**43**(4):237-246. DOI: 10.1016/S0013-7952(96)00033-6

[34] Ayalew L, Yamagishi H, Ugawa N. Landslide susceptibility mapping using GIS-based weighted linear combination, the case in Tsugawa area of Agano River, Niigata prefecture, Japan. Landslides. 2004;**1**(1):73-81. DOI: 10.1007/s10346-003-0006-9

[35] Wati SE, Hastuti T, Widjojo S, Pinem F. Landslide susceptibility mapping with heuristic approach in mountainous area: A case study in Tawangmangu sub-district Central Java, Indonesia. The International Archives of the Photogrammetry, Remote Sensing and Spatial Information Sciences. 2010;**38**(Part 8):249-252

[36] Bijukchhen SM, Kayastha P, Dhital MR. A comparative evaluation of heuristic and bivariate statistical modelling for landslide susceptibility mappings in Ghurmi–Dhad Khola, East Nepal. Arabian Journal of Geosciences. 2013;**6**(8):2727-2743. DOI: 10.1007/s12517-012-0569-7

[37] Carrara A, Cardinali M, Guzzetti F, Reichenbach P. GIS technology in mapping landslide hazard. In: Carrara A, Guzzetti F, editors. Geographical Information Systems in Assessing Natural Hazards. Advances in Natural and Technological Hazards Research. 5th ed. Dordrecht: Springer; 1995. pp. 135-175. DOI: 10.1007/978-94-015-8404-3_8

[38] Irigaray C. Cartografía de Riesgos Geológicos asociados a los movimientos de ladera en el sector de Colmenar (Málaga) [thesis]. Granada: University of Granada; 1990

[39] Irigaray C. Movimientos de ladera: inventario, análisis y cartografía de susceptibilidad mediante la aplicación de un Sistema de Información Geográfica. Aplicación a las zonas de Colmenar (Málaga), Rute (Córdoba) y Montefrío (Granada) [thesis]. Granada: Departamento de Ingeniería Civil de la Universidad de Granada; 1995

[40] Van Westen CJ. Medium scale landslide hazard analysis using a PC based GIS. A case study from Chinchina,

Colombia. In: 1er. Simposio Internacional sobre Sensores Remotos y Sistemas de Información Geográfica (SIG) para el estudio de riesgos naturales; Bogotá, Colombia. 1992. pp. 1-15

[41] Van Westen CJ. Application of Geographic Information Systems to Landslide Hazard Zonation. The Netherlands: ITC; 1993. ISBN: 90-6164-078-4

[42] Sarkar S, Kanungo DP, Mehrotra GS. Landslide hazard zonation: A case study in Garhwal Himalaya, India. Mountain Research and Development. 1995;**15**(4):301-309. DOI: 10.2307/3673806

[43] Barredo JI, Benavides A, Hervás J, van Westen CJ. Comparing heuristic landslide hazard assessment techniques using GIS in the Tirajana basin, gran Canaria Island, Spain. International Journal of Applied Earth Observation and Geoinformation. 2000;**2**(1):9-23. DOI: 10.1016/S0303-2434(00)85022-9

[44] Hervás J, Barredo JI. Evaluación De La Susceptibilidad De Deslizamientos Mediante El Uso Conjunto De SIG Teledetección y Métodos De Evaluación Multicriterio. Aplicación al barranco de Tirajana (Gran Canaria). In: V Simposio nacional sobre taludes y laderas Inestables. 2001

[45] Jenks GF, Caspall FC. Error on choroplethic maps: Definition, measurement, reduction. Annals of the Association of American Geographers. 1971;**61**(2):217-244. DOI: 10.1111/j.1467-8306.1971.tb00779

[46] INEGI Monografía del Estado de Guerrero; 2013. p. 37

[47] Precipitación. Available from: https://smn.conagua.gob.mx/es/climatologia/pronostico-climatico/precipitacion-form [Accessed: 06 April 2019]

[48] INEGI. Guía para la interpretación de cartografía: uso del suelo y vegetación, escala 1:250, 000, serie V. Cartografía - Estudio y enseñanza. ed. México: Instituto Nacional de Estadística y Geografía; 2015

[49] INEGI. Censo de Población y Vivienda 2010. México: Resultados definitivos; 2011

[50] Raisz E. Landforms of Mexico: Cambridge, Massachusetts. US Navy Geographical Branch, mapa, escala 1:3,000,000; 1964

[51] Cerca-Martínez M. Deformación y magmatismo Cretácico Tardío-Terciario Temprano en la zona de la Plataforma Guerrero-Morelos [thesis]. México DF: Universidad Nacional Autónoma de México; 2004

[52] Pérez-Gutiérrez R. Geología y evolución estructural del Complejo Xolapa, entre los ríos Papagayo y La Sabana, noreste de Acapulco, Guerrero, México [thesis]. México: Universidad Nacional Autónoma de México; 2005

[53] Ramos-Bernal R, Vázquez-Jiménez R, Romero-Calcerrada R, Arrogante-Funes P, Novillo C. Evaluation of unsupervised change detection methods applied to landslide inventory mapping using ASTER imagery. Remote Sensing. 2018;**10**(12):1987. DOI: 10.3390/rs10121987

[54] Figueroa H. Huracán Ingrid y tormenta Manuel inundan al país [Internet]. 2013. Available from: https://bit.ly/2PniA2L [Accessed: 12 July 2019]

[55] Alcantara-Ayala I. Índice de susceptibilidad a movimientos del terreno y su aplicación en una región semiárida. Revista Mexicana de Ciencias Geológicas. 2000;**17**(1):66-75

[56] Ramos-Bernal RN. Estudio de la susceptibilidad al deslizamiento

de laderas en el Estado de Guerrero, México, aplicando Tecnologías de Información Geográfica [thesis]. Spain: King Juan Carlos University; 2018

[57] Aramburu Maqua MP, Escribano BR. Guía para la elaboración de estudios del medio físico. 3rd ed. Madrid: Centro de Publicaciones. Secretaría General Técnica. Ministerio de Medio Ambiente; 2006. ISBN: 84-8320-374-X

[58] Silverman BW. Using kernel density estimates to investigate multimodality. Journal of the Royal Statistical Society: Series B: Methodological. 1981;**43**:97-99

[59] Silverman BW. Estimación de densidad para las estadísticas y el análisis de datos. New York: Chapman and Hall; 1986

[60] Rawat MS, Uniyal DP, Dobhal R, Joshi V, Rawat BS, Bartwal A, et al. Study of landslide hazard zonation in Mandakini Valley, Rudraprayag district, Uttarakhand using remote sensing and GIS. Current Science. 2015;**109**(1):158-170

[61] Pourghasemi HR, Moradi HR, Aghda SMF. Landslide susceptibility mapping by binary logistic regression, analytical hierarchy process, and statistical index models and assessment of their performances. Natural Hazards. 2013;**69**(1):749-779. DOI: 10.1007/s11069-013-0728-5

[62] Ohlmacher GC. Plan curvature and landslide probability in regions dominated by earth flows and earth slides. Engineering Geology. 2007;**91**(2):117-134. DOI: 10.1016/j.enggeo.2007.01.005

[63] Convertino M, Troccoli A, Catani F. Detecting fingerprints of landslide drivers: A MaxEnt model. Journal of Geophysical Research - Earth Surface. 2013;**118**(3):1367-1386. DOI: 10.1002/jgrf.20099

[64] O'Banion MS, Olsen MJ. Predictive seismically-induced landslide hazard mapping in Oregon using a maximum entropy model (MaxEnt). In: Proceedings of the 10th National Conference in Earthquake Engineering, Earthquake Engineering Research Institute, Anchorage. Anchorage, AK: Earthquake Engineering Research Institute; 2014. p. 2014

[65] Lombardo L, Fubelli G, Amato G, Bonasera M. Presence-only approach to assess landslide triggering-thickness susceptibility: A test for the Mili catchment [North-Eastern Sicily, Italy]. Natural Hazards. 2016;**84**(1):565-588. DOI: 10.1007/s11069-016-2443-5

[66] Kornejady A, Ownegh M, Bahremand A. Landslide susceptibility assessment using maximum entropy model with two different data sampling methods. Catena. 2017;**152**:144-162. DOI: 10.1016/j.catena.2017.01.010

[67] Soares-Filho BS, Cerqueira GC, Pennachin CL. DINAMICA—A stochastic cellular automata model designed to simulate the landscape dynamics in an Amazonian colonization frontier. Ecological Modelling. 2002;**154**(3):217-235. DOI: 10.1016/S0304-3800(02)00059-5

[68] Soares-Filho BS, Rodrigues HO, Costa W. Handbook of Modeling Environmental Dynamics with Dinamica EGO. Belo Horizonte, Minas Gerais: Centro De Sensoriamento Remoto. Universidade Federal de Minas Gerais; 2009. p. 115. ISBN: 978-85-910119-0-2

[69] CSR-UFMG. What Is Dinamica EGO? [Internet]. 2015. Available from: http://csr.ufmg.br/dinamica/ [Accessed: 11 July 2019]

[70] Recondo C, Menéndez C, García P, González R, Sáez E. Estudio de las zonas propensas a sufrir deslizamientos en los Concejos de Oviedo y Mieres [Asturias]

a partir de una imagen Landsat-TM y de un Modelo Digital de Elevaciones. Revista de Teledetección. 2000;**14**:49-59

[71] Pourghasemi HR, Mohammady M, Pradhan B. Landslide susceptibility mapping using index of entropy and conditional probability models in GIS: Safarood Basin, Iran. Catena. 2012a;**97**:71-84. DOI: 10.1016/j.catena.2012.05.005

[72] Pradhan B, Lee S. Delineation of landslide hazard areas on Penang Island, Malaysia, by using frequency ratio, logistic regression, and artificial neural network models. Environment and Earth Science. 2010a;**60**(5):1037-1054. DOI: 10.1007/s12665-009-0245-8

[73] Pourghasemi HR, Pradhan B, Gokceoglu C. Application of fuzzy logic and analytical hierarchy process [AHP] to landslide susceptibility mapping at Haraz watershed, Iran. Natural Hazards. 2012b;**63**(2):965-996. DOI: 10.1007/s11069-012-0217-2

[74] Egan JP. Handbook of Signal Detection Theory and ROC Analysis. London, UK: Academic Press; 1975

[75] Yesilnacar EK. The Application of Computational Intelligence to Landslide Susceptibility Mapping in Turkey [Thesis]. University of Melbourne, Department, 200; 2005

[76] Pradhan B, Lee S. Landslide susceptibility assessment and factor effect analysis: Backpropagation artificial neural networks and their comparison with frequency ratio and bivariate logistic regression modelling. Environmental Modelling and Software. 2010b;**25**(6):747-759. DOI: 10.1016/j.envsoft.2009.10.016

[77] Chen W, Pourghasemi HR, Kornejady A, Zhang N. Landslide spatial modeling: Introducing new ensembles of ANN, MaxEnt, and SVM machine learning techniques. Geoderma. 2017;**305**:314-327. DOI: 10.1016/j.geoderma.2017.06.020

[78] Devkota KC, Regmi AD, Pourghasemi HR, Yoshida K, Pradhan B, Ryu IC, et al. Natural Hazards. 2013;**65**(1):135-165. DOI: 10.1007/s11069-012-0347-6

[79] Kim HG, Lee DK, Park C, Kil S, Son Y, Park JH. Evaluating landslide hazards using RCP 4.5 and 8.5 scenarios. Environment and Earth Science. 2015;**73**(3):1385-1400. DOI: 10.1007/s12665-014-3775-7

[80] Clerici A, Perego S, Tellini C, Vescovi P. A procedure for landslide susceptibility zonation by the conditional analysis method. Geomorphology. 2002;**48**(4):349-364. DOI: 10.1016/S0169-555X(02)00079-X

Chapter 5

Modeling Antecedent Soil Moisture to Constrain Rainfall Thresholds for Shallow Landslides Occurrence

Maurizio Lazzari, Marco Piccarreta, Ram L. Ray and Salvatore Manfreda

Abstract

Rainfall-triggered shallow landslide events have caused losses of human lives and millions of euros in damage to property in all parts of the world. The need to prevent such hazards combined with the difficulty of describing the geomorphological processes over regional scales led to the adoption of empirical rainfall thresholds derived from records of rainfall events triggering landslides. These rainfall intensity thresholds are generally computed, assuming that all events are not influenced by antecedent soil moisture conditions. Nevertheless, it is expected that antecedent soil moisture conditions may provide critical support for the correct definition of the triggering conditions. Therefore, we explored the role of antecedent soil moisture on critical rainfall intensity-duration thresholds to evaluate the possibility of modifying or improving traditional approaches. The study was carried out using 326 landslide events that occurred in the last 18 years in the Basilicata region (southern Italy). Besides the ordinary data (i.e., rainstorm intensity and duration), we also derived the antecedent soil moisture conditions using a parsimonious hydrological model. These data have been used to derive the rainfall intensity thresholds conditional on the antecedent saturation of soil quantifying the impact of such parameters on rainfall thresholds.

Keywords: landslides, soil saturation, geomorphology, hydrogeological risk, Basilicata

1. Introduction

Rainfall-induced shallow landslides are critical issues of scientific and societal interest, causing billions of euros in damages and thousands of deaths every year [1]. A large number of studies investigated the functional relationship between rainfall characteristics and landslide events [2]. One of the main results is the definition of empirical rainfall thresholds associated with the triggering of the shallow landslide, such as total event rainfall, intensity-duration, event-duration, and event-intensity thresholds ([3] and reference therein) [4, 5]. However, these approaches lead to a limited understanding of the geomorphological process and, if used for warning purposes, they can produce a large number of false positives alarms [6]. In fact, rainfall thresholds approach evaluates only the amount of cumulated rainfall and it neglects the primary role of other vital parameters, such as evapotranspiration, soil moisture, rainfall infiltration, soil porosity, and permeability.

In order to consider predisposing hydrological factors on empirical threshold calculation, recent studies have focused on the role of the antecedent daily rainfall in landslides triggering [7–12]. These approaches have found a strong relationship between the hourly rainfall data triggering landslides and the initial soil moisture contributing to improving the predictive accuracy of empirical thresholds. Those results have also stimulated a critical revision of the intensity/duration thresholds in the last few years [6, 13–15]. In particular, Bogaard and Greco [6] introduced the cause-trigger concept for defining hydro-regional thresholds for predicting landslide occurrence, also suggesting taking into consideration the slope water balance. Starting from this new perspective, we aim to contribute to this discussion by evaluating the correlation between antecedent soil moisture conditions and rainfall intensity during shallow landslide events. In particular, we would like to explore better how much the initial saturation degree of soil affects the intensity/duration (I/D) relationships in landslide prediction. For this purpose, it is very important to use reliable databases in the literature or otherwise build a specific one.

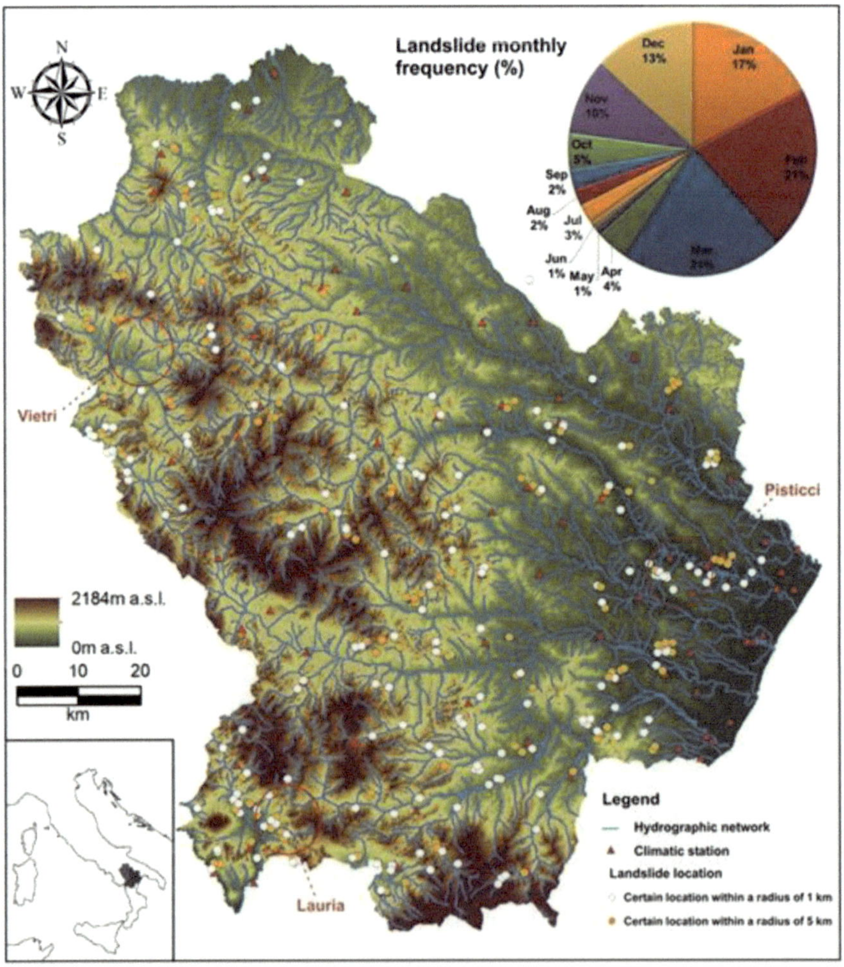

Figure 1.
Geographical distribution of the weather stations and landslide events for the study area. The graph in the inset shows the monthly distribution of landslides in Basilicata from 2001 to 2018.

There are many soil moisture datasets that have been successfully used to calibrate and validate catchment or watershed scale models of infiltration, soil-moisture storage, and in some ways, even to examine the first landslide trigger [16–19].

In this chapter, we addressed this issue by reconstructing and leveraging a dataset of 326 landslide events that occurred in the Basilicata region (southern Italy, **Figure 1**, and Appendix), from January 2001 to March 2018. For each georeferenced landslide, we derived the rainfall event characteristics and antecedent soil moisture conditions using a parsimonious physically-based distributed model applied at the regional scale with a spatial resolution of 200 m and a daily and hourly time scale. This approach allowed us to reconstruct all of the main forcing factors that may have produced a change in the slope stability and to detect the impact of antecedent soil moisture on the rainfall intensity/duration relationship. The numerical simulation has been needed to reconstruct the antecedent soil moisture values not available for the whole study area.

2. Data and methods

This study was carried out within a research agreement with the Civil Protection of the Basilicata region, which supported in part the reconstruction of the list of landslide events used for the analysis. The database was constructed with the primary aim of creating an updated description of the most recent landslides and the associated rainfall events. Therefore, the present section will be devoted to the description of the study area, the methodology adopted to build the database, and the modeling approach used to reconstruct the antecedent soil saturation conditions.

2.1 Study area

Basilicata is a region of southern Italy covering an area of 9.992 km^2 characterized by different topographical and geomorphological contexts, landscape types (47% mountains, 45% hillocks, and 8% plains) and geolithological conditions. The north-western and south-western regions are characterized by mountain landscapes (southern Apennines) with significant elevations of the relief (between 1300 and 2000 m of altitude) and steep slopes, particularly where Mesozoic successions (dolomite and siliceous limestones) outcrop. The eastern region shows a hilly landscape characterized by soft shapes or tabular hills (alternating ridges and valleys in conglomeratic sandstone—clayey—marly), usually with low gradients of the slopes, often modeled in foredeep Plio—Pleistocene units with clayey dominant [20, 21].

Precipitation values are typical of the Mediterranean, with distinct dry and wet seasons [22]. Higher precipitation totals occur during the last autumn-winter period when landslides and floods usually take place (more than 70%). A near real-time hydrometeorological network covers the territory uniformly with a density of one station every 80 km^2. It has been operating over a time interval of about 70 years, providing temperature and precipitation data at the resolution of 10 minutes.

2.2 Landslide and rainfall data

Based on detailed bibliographical research [23–25], which explored all available sources including national and local newspapers and journals, Internet blogs, and the scientific and technical literature, we have collected a database of 326 shallow landslide events (landslide event is a single landslide) from January 2001 to March 2018.

The information collected and stored in the inventory includes (**Figure 1**):

- accurate or approximate location of the landslide event;
- accurate or approximate time, date, or period of the failures;
- rainfall conditions that resulted in slope failures collected from the nearest rain gauge, including the total event rainfall, the rainfall duration, the mean rainfall intensity, and the antecedent rainfall for 2001–2018;
- landslide type;
- a generic description of the lithology.

In addition to this data that is also reported in Appendix in a tabular format, meteorological data and the output of the hydrological model have been used for the subsequent elaborations. In particular, hourly rainfall and temperature data were obtained from the rain gauges of the Civil Protection of the region. The hydrological model proposed on a regional scale considers homogeneous soil moisture conditions in the space and the first meters of depth in the areas affected by each landslide identified in our database.

The regional pedological map is depicted in **Figure 2** with the spatial distribution of landslides (**Figure 2**). This maps provides a nested description of soil classes that indentifies four regions at the first level (soil map of Italy, scale 1: 5,000,000), the 15 provinces, and 75 soil units (scale 1: 250,000). Based on the pedological characteristics of the regions, it was observed that the highest number of landslides (56 landslides) occurred in the soil province n.6. It is also worthy to mention a high number of events occurred on the soil unit 12.4 (33 landslides) and 10.2 (21 landslides). These last two soil units correspond to:

- 12.4—hilly clay soils with steep slopes, badlands, intended for grazing or arable land, with low permeability (Vertic Haploxerepts; Inceptisol);
- 10.2—hilly sandy-conglomerate soils, intended for pasture, vineyards or shrubs (Typic Xerorthents; Inceptisol).

The number of event recorded in each soil unit is described in the histogram of **Figure 3**.

2.3 Reconstruction of rainfall events

The rainfall duration (D) was determined by measuring the time between the moment of the beginning of each rainfall event, which triggered a shallow landslide, considered in the database, and rainfalls ending time. The rainfall ending time was taken to coincide with the time of the last rainfall measurement of the day when the landslide occurred. As suggested by Brunetti et al. [26], the starting time was considered a minimum period without rain (a 2-day period without rainfall was selected for late spring and summer, May–September, and a 4-day period without rainfall was selected for the other seasons, October–April) to separate two consecutive rainfall events. Once the duration of the rainfall event was established, the corresponding rainfall mean intensity I (mm h^{-1}) was calculated dividing the cumulated (total) rainfall (mm) in the considered period by the length of the rainfall period (hours). The full list of events is given in the Appendix of the present

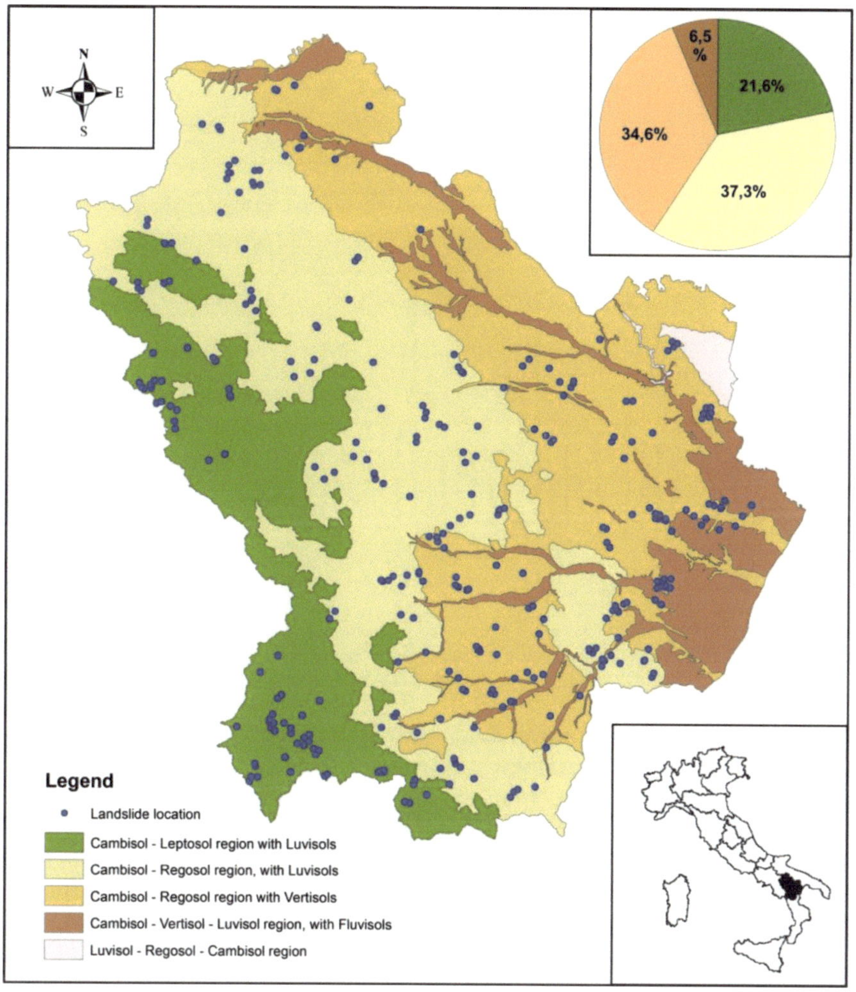

Figure 2.
Pedological regions and shallow landslides distribution over the Basilicata region.

chapter (**Table 1**—Appendix). In recent studies, authors [27–29] have used an approach that provides the seasonality criterion (April-October for the "dry/warm" season and November-March for the "wet/cold" season) to calculate the rainfall events. In this chapter, the proposed method is different from that proposed by Peruccacci et al. [29], because the saturation value and condition are a parameter regardless of seasonality. It provides a more detailed parameter, overcoming the possibility that in the same season, it can have more dry or wet phases.

2.4 Modeling soil water content

Although antecedent soil moisture can be obtained by in-situ measurements at a point scale, measurements on a regional scale are time-consuming and expensive. Recently, more information is available from satellite data, but they are too coarse to provide local estimates of soil water content on a specific landslide [30]. Thus, we used the hydrological model AD2 to describe the temporal evolution of soil water content over the entire Basilicata region using a distributed approach at 240 m spatial resolution.

The AD2 model is a 1D model capable of describing the soil water budget along the vertical direction, but its physically based nature allows to associate physical

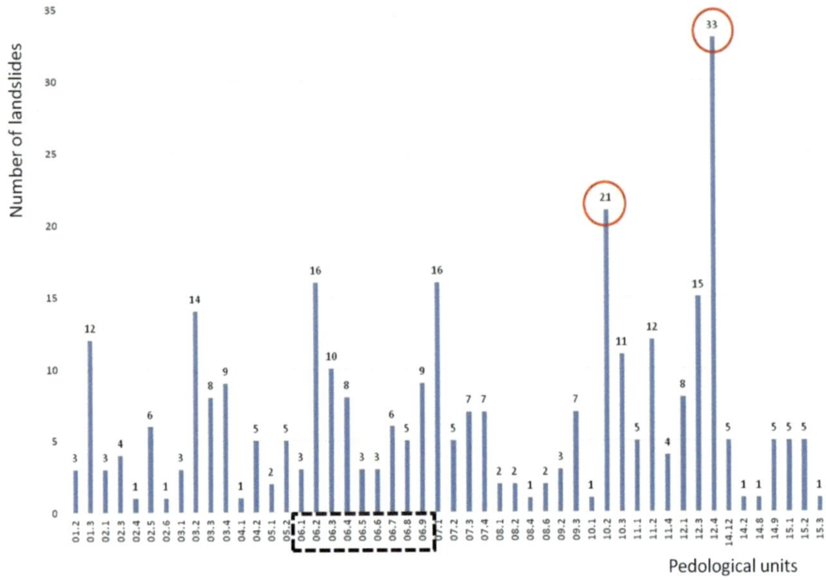

Figure 3.
Histogram with the distribution of the number of landslides in the various regional soil units. Red circles, units with multiple landslides; black dotted rectangle, landslides included in unit 6.

characteristics such as soil texture, land cover, and mean slope to each pixel/location that affects the model parametrization.

The model parameters were obtained from physical maps such as national and regional pedological maps of Italy and Basilicata [31], the III level of the CORINE Land Cover map [32], and the Shuttle Radar Topography Mission Digital Elevation Model (SRTM-DEM) extracted from HydroSHEDS (hydrosheds.cr.usgs.gov/index.php).

The model was run at 1 h temporal resolution using rainfall and temperature data derived from the rainfall network of the Civil Protection for the period January 1, 2001 to March 28, 2018 [32].

This approach is straightforward and can be easily replicated elsewhere after a simple calibration against the local soil moisture and landslide datasets. It must be stated that obtained values can be affected by several errors due to model structure, parametrization, and climatic data, but at present, such an approach offers a realistic description of the expected relative saturation of the soil providing a synthesis of the state of the system according to the available information on soil texture, antecedent rainfall, and evolution of temperatures. Moreover, several evidences suggesting that the use of a physically based approach allows obtaining more robust outputs [33, 34].

2.4.1 AD2 model structure

Model simulations carried out using at least 1 year of rainfall and temperature data recorded before each landslide event to reach a reliable estimate of the relative soil water content at the date of the considered event. AD2 [34] provides a hydrological prediction that considers several hydrological components such as infiltration, surface runoff, sub-surface runoff, deep percolation, and evapotranspiration. Soil water balance is described by the following Equation [35]:

$$S_{t+\Delta t} = S_t + I_t - R_{out,t} - L_t - E_t, \quad (1)$$

where: S_t is the basin soil water content at the generic instant of time t, which represents a key variable of the model influencing runoff production, leakage, and evapotranspiration; I_t is the infiltration; $R_{out,t}$ is the sub-surface runoff production; L_t is the leakage to the groundwater; and E_T is the actual evapotranspiration.

The infiltration is derived from the difference between the rainfall amount, P_t, and the surface runoff, R_t, at time t (mm):

$$I_t = P_t - R_t. \tag{2}$$

Runoff is calculated using the equation proposed by De Smedt et al. [36], which takes into account the potential saturation of the soil:

$$R_t = \begin{cases} \left(\frac{S_t}{S_{max}}\right) P_t \text{ if } P_t \leq P_c = \frac{S_{max}(S_{max} - S_t)}{(S_{max} - CS_t)} \\ P_t - (S_{max} - S_t) \text{ if } P_t > P_c = \frac{S_{max}(S_{max} - S_t)}{(S_{max} - CS_t)} \end{cases} \tag{3}$$

where, S_{max} is the maximum water storage capacity of the bucket, P_c is the critical rainfall producing the surface soil saturation, and C the default runoff coefficient that is parameterized as a function of soil type, soil cover, and slope [37].

The sub-surface runoff production is assumed to be a linear function of the soil water content above the field capacity reference parameter:

$$R_{(out,t)} = \max\{0, c(S_t - S_c)\}, \tag{4}$$

where S_c is the threshold water content for sub-surface flow production, assumed here equal to 0.6 S_{max}, and c is the sub-surface coefficient, which is generally assumed 0.05.

The evapotranspiration is assumed to be a bi-linear function of the soil content and potential evapotranspiration. It may be described by the following equation:

$$E_t = \max\left\{0, \min\left\{\left(\frac{S_t}{0.75S_c}\right) EP, EP\right\}\right\}, \tag{5}$$

where E_P is the potential evapotranspiration, $0.75S_c$ is an estimate of the water content at which the stomata closure starts to reduce the evapotranspiration.

Leakage is computed using the expression derived by Manfreda et al. [38] integrating the power-law function of leakage by Eagleson [39] over a time-step Δ_t:

$$L_t = \begin{cases} 0 & \text{if } S_t \leq S_c \\ \left(S_{t-1} - S_{max}\left(\frac{\Delta t K_s}{S_{max}} + \left(\frac{S_{t-1}}{S_{max}}\right)^{1-\beta}\right)^{1/(1-\beta)}\right) & \text{if } S_t > S_c \end{cases}, \tag{6}$$

where L_t is the groundwater recharge in Δ_t, K_s is a parameter that interprets the soil permeability at saturation, and β is a dimensionless exponent.

It must be clarified that all the parameters mentioned in the model equations reported above can be estimated using the existing literature values that associate this parameter to physical features of the area such as soil texture, land use and mean slope using [37, 40, 41].

2.5 Rainfall thresholds

To determine rainfall thresholds for shallow landslide occurrence, we adopted the Frequentist method [26]. The threshold curve is assumed to follow a power law:

$$I = \alpha D^{-\beta} \tag{7}$$

where, I is the rainfall mean intensity (mm h^{-1}), D is the rainfall event duration (h), α is the intercept, and β defines the slope of the power law function. Empirical data were log-transformed to calculate the best-fit line by means of a linear equation $\log(I) = \log(\alpha) - \beta \log(D)$, equivalent to that described above.

Following the methods adopted in previous studies [9, 12, 15], we identified the rainfall events associated with each landslide event and the corresponding degree of soil saturation at the starting time of each event. Including this additional information in the database, it was possible to explore its role in the general behavior of the rainfall events triggering landslides under different initial conditions.

3. Results and discussion

The comparison between the rainfall intensity and the relative saturation before each event is depicted in **Figure 4**. **Figure 4** provides the temporal evolution of the rainfall and relative soil saturation of three different sites (Lauria, Vietri di Potenza and Pisticci; see Appendix) characterized by different lithological conditions during the period from January 1, 2009 to December 31, 2015. This window was extracted from the model simulation to emphasize the seasonal dynamics of soil moisture over the considered sites. Such a seasonality is clearly one of the motivations to conduct this study because such a dynamic strongly affects the hydraulic processes in the soil profile.

It is noticed that most of the different landslide events (reported in the graph with a dark star) occurred after significant rainfall amounts and relatively high or moderate soil saturation degree. When the same rainfall amounts occurred in conditions of low antecedent soil moisture content, they have not produced shallow landslides. This finding aligns with the previous studies [7–15, 42].

This preliminary plot shows that there is an interplay between the antecedent soil moisture conditions and the amounts and the duration of the triggering rainfall. Moreover, it appears how the same degree of soil saturation and rainfall I/D conditions do not necessarily produce the same effects in different geopedological regions.

The role of the antecedent soil moisture condition on the triggering rainfall intensity is clearly shown in **Figure 5**, where the rainfall intensity/duration has been plotted against the simulated antecedent soil saturation of each landslide event. This graph was developed following the trigger-cause concept of Bogaard and Greco [6] and highlights the role of both rainfall dynamics and antecedent soil moisture on the slope stability.

In fact, there is a clear reduction of the rainfall intensity needed to trigger a landslide with the increase of the antecedent soil moisture. In this graph, the data grouped in the function of the rainfall duration trying to explore also the role of this additional parameter on the process. It is observed that the rainfall dynamics also matter, being shorter rainfall events more sensitive to the antecedent soil moisture respect to, while more extended events are less influenced by such parameter.

Similarly, previous studies [7, 8, 12, 43] also found a linearly decreasing trend between the mean rainfall intensity and the initial soil moisture conditions. The slope of the regression functions derived from a different subset of our database changes based on the relative duration of the rainfall events. It is higher for rainfall durations lower than 48 h, while the function becomes almost independent from the relative saturation when rainfall events have longer durations (more than 48 h). This is probably due to the nature of the long-lasting rain events, which are often characterized by a high total amount of rainfall. Results in high values both of the initial saturation degree and of low rainfall intensity that is averaged over longer periods. It must be clarified that the relative degree of saturation has been referred to as the starting time of the triggering rainfall event.

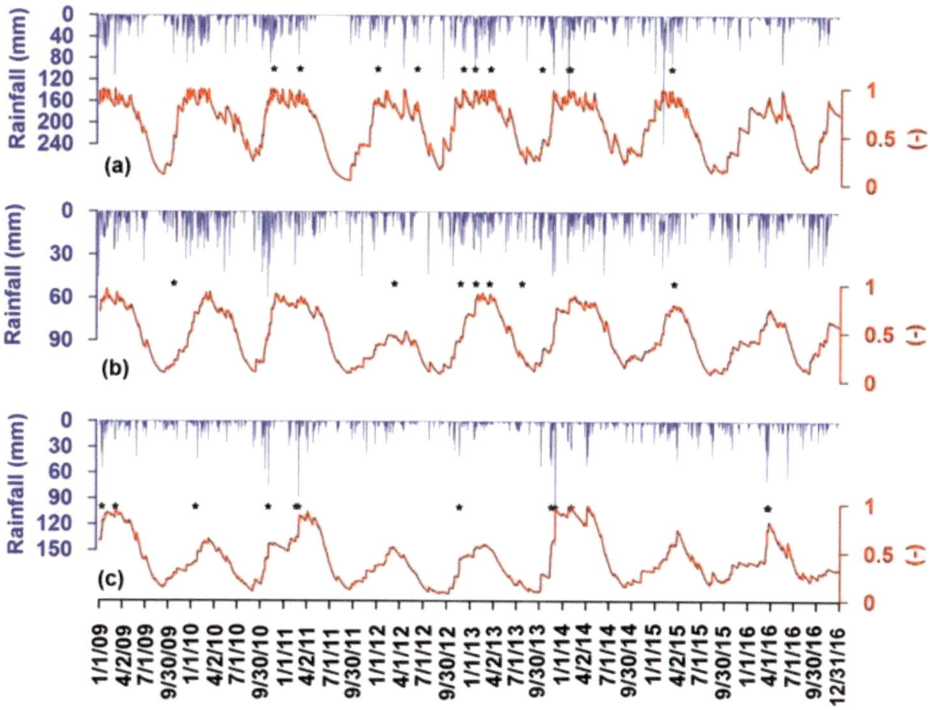

Figure 4.
Daily rainfall (blue) and simulated daily soil degree saturation (red) at (a) Lauria, (b) Vietri, and (c) Pisticci from January 1, 2009 to December 31, 2016. Dark stars represent the data of the occurrence of shallow landslide events in the monitored areas.

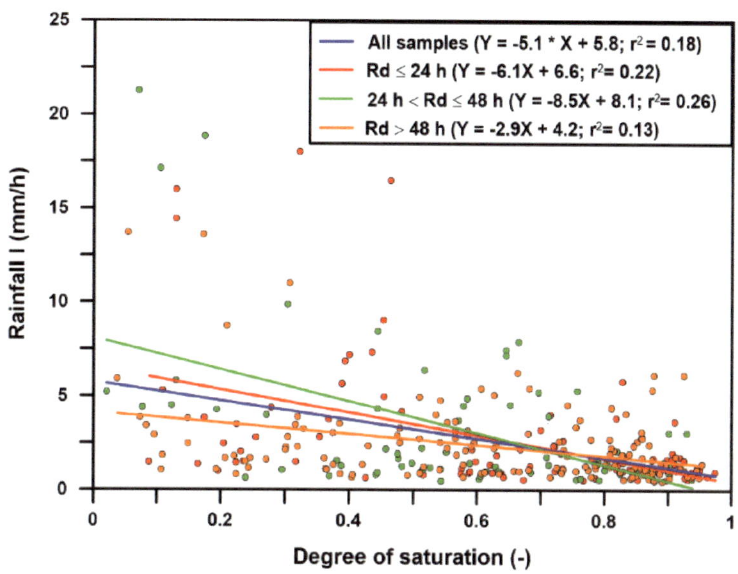

Figure 5.
Mean rainfall intensity/duration and the simulated initial degree of saturation for the 326 landslide events in Basilicata region (southern Italy) from 2001 to 2018.

Figure 6 depicts a clear picture of the dependence between mean rainfall intensity of event of different durations and the antecedent soil moisture, where the rainfall intensity values of the 326 investigated events are associated to the simulated soil saturation degree using a color scale (from blue to yellow starting from lower to higher values of degree of saturation). This graph clearly shows that higher amounts of rainfall intensity are observed in correspondence to lower values of soil saturation and vice-versa. This tendency is not always consistent due to the presence of several spurious data relative to the occurrence of extraordinarily wet events, which resulted in both landslides and floods.

To evaluate the role of degree of soil saturation on the regional mean rainfall intensity/duration function, we have identified two distinguished sub-samples based on the antecedent soil moisture conditions of each event. The two groups were distinguished using a sensitivity analysis, exploiting different antecedent soil saturation values. The selection was made using a subjective selection that tried to identify the most diverse groups of landslides using a given threshold of soil saturation. Therefore, we determined mean rainfall intensity/duration functions (rainfall thresholds) under middle-low antecedent soil moisture conditions that seemed to those that responded better to the data considered (soil degree saturation lower than 0.70), and moderate to high antecedent soil moisture conditions (soil degree saturation equal or higher than 0.70). In this way, it was possible to derive critical rainfall threshold functions conditional on the antecedent soil moisture conditions.

The two functions plotted in the graph (**Figure 6**), which has significantly different slopes. This implies that they must cross somewhere in the space of rainfall intensities and event duration. In the present case, we observed that they cross in a point corresponding to the duration of about 200 h. At such duration, the

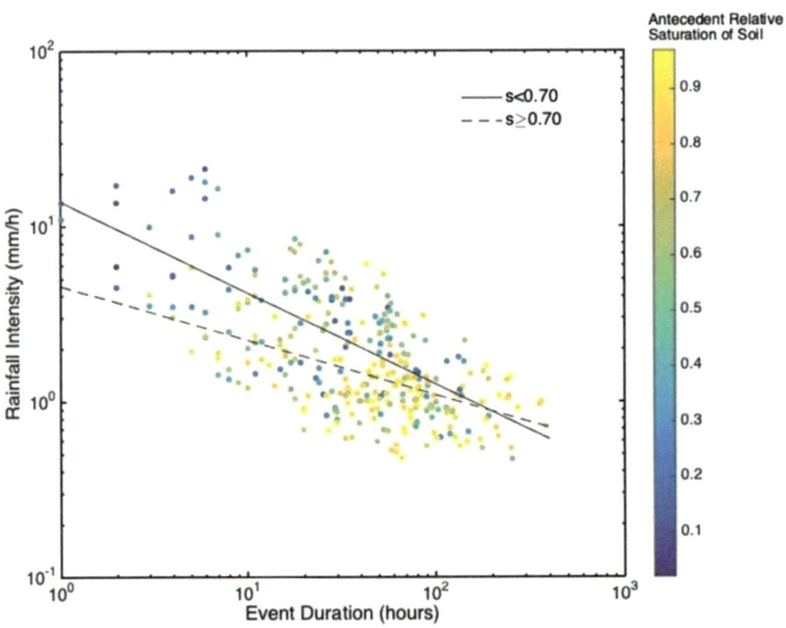

Figure 6.
Rainfall intensity as a function of the duration of the triggering rainfall events for the 326 landslide events recorded in Basilicata region (southern Italy) during the period 2001–2018. Each event is associated with a color that represents the simulated antecedent degree of saturation, whose range is given in the color bar on the right (ranging from 0 to 1). We also included the regression lines estimated for the two groups of events selected based on the antecedent soil saturation conditions. The solid line represents the regression function obtained using the observations with soil degree saturation lower than 0.70, while the dotted line represents the regression function obtained using the observations with degree soil of saturation equal or higher than 0.70.

impact of antecedent soil water content becomes not relevant, and this part of the curve should not be considered.

The proposed approach allows taking into account both rainfall characteristics (intensity and duration) and the antecedent soil moisture state in a specific study area, contributing to foresee a landslide event.

Of course, a methodology like this should be evaluated widely, also taking into consideration the ability of the method to distinguish between true and false alarms. Unfortunately, this field-test is challenging to be implemented in a region like Basilicata with a low density of population (as single possible observators), where a lot of landslide events are not reported or are missing.

4. Final remarks

We have explored the role and effects of antecedent soil moisture conditions on rainfall I/D thresholds triggering shallow landslides by using a dataset built for a region of southern Italy and a distributed modeling approach. By combining rainfall events data with the simulated antecedent soil moisture conditions, it was possible to derive I/D relationships, which can be used to discriminate the triggering conditions for landslides better.

Two distinct degree of soil saturation values [$S < 0.7$ and $S \geq 0.7$] were identified to distinguish different classes of events. Such soil moisture conditions led to two distinct populations of events that identified statistically significantly different rainfall threshold functions. Our results are consistent with those found in the most recent studies on this topic, reinforcing the idea that simulated soil moisture provides better metrics than antecedent rainfall for the predisposing factors of landslide initiation.

Finally, a forthcoming extension of this research will aim to carry out a local downscaling to define the relations between I/D and the degree of soil saturation in the smallest territorial contexts characterized by the same climatic and lithotechnical conditions, in which the landslides inserted in our database have developed.

Moreover, it is also important to note that the proposed description of the landslide event may undoubtedly support the development of further studies and models for landslide prediction. In fact, the main results obtained in the present study are the fact that the information about the antecedent relative saturation of the soil may help to distinguish the dynamics of the process better. Therefore, it would be a good practice to include such parameters in all landslide database. This can happen with the support of remote sensing techniques that also allow deriving root zone soil moisture over large areas [10, 41, 44].

Acknowledgements

This work was carried out within a scientific agreement between the Civil Protection Department of Basilicata, the Interuniversity Consortium for Hydrology (CINID), and the University of Basilicata to the start-up the Basilicata Hydrologic Risk Center.

Conflict of interest

The authors declare no conflict of interest.

Appendix

ID	Municipality	Date	UTM N	UTM E	H (mm)	D (h)	I (mm/h)	Antecedent soil saturation	Weather station	Sources
1	Rotondella	14/01/2001	4,447,838,6	630,012,0	186,8	26,0	7,18	0,40	Nova Siri SAL	Evalmet web site web site web site
2	Montalbano Jonico	20/01/2001	4,461,333,2	632,500,7	66,6	16,0	4,16	0,48	Montalbano SAL	Civil Protection
3	Pisticci	20/01/2001	4,472,103,8	633,501,3	83,6	34,0	2,46	0,20	Pisticci Scalo SAL	Evalmet web site
4	Rotondella	21/01/2001	4,447,722,3	629,857,3	88,6	52,0	1,70	0,78	Nova Siri Sal	Evalmet web site
5	Tursi	21/01/2001	4,456,520,6	624,987,7	86,2	27,0	3,19	0,70	Tursi SI	Evalmet web site
6	San Fele	09/03/2002	4,518,711,1	545,705,9	23,6	11,0	2,15	0,59	San Fele PC	Civil Protection
7	Lagonegro	12/01/2003	4,441,050,6	565,797,2	148,2	140,0	1,06	0,89	Lagonegro PC	Civil Protection
8	Acerenza	25/01/2003	4,516,566,8	579,515,5	43,0	43,0	1,00	0,20	Acerenza SAL	Civil Protection
9	Nova Siri	25/01/2003	4,444,982,6	631,308,4	80,0	56,0	1,43	0,91	Nova Siri SAL	Civil Protection
10	Pisticci	25/01/2003	4,472,200,4	631,782,9	54,2	48,0	1,13	0,82	Pisticci Scalo SAL	La Nuova Basilicata magazine
11	Castronuovo di Sant'Andrea	04/02/2003	4,449,625,0	600,976,5	33,4	34,0	0,98	0,90	Roccanova PC	Civil Protection
12	Muro Lucano	08/02/2003	4,512,107,1	540,718,1	105,6	172,0	0,61	0,87	Muro Lucano PC	Civil Protection
13	Montescaglioso	09/09/2003	4,490,159,8	640,165,8	21,2	4,0	5,30	0,11	Montescaglioso SAL	Civil Protection
14	Venosa	10/09/2003	4,535,319,9	569,252,5	35,0	8,0	4,38	0,23	Venosa SAL	Civil Protection
15	Craco	12/12/2003	4,467,134,0	623,760,9	117,6	78,0	1,51	0,24	Craco PC	Piccarreta et al., 2004
16	Montescaglioso	26/07/2004	4,490,946,3	640,686,5	31,4	23,0	1,37	0,16	Montescaglioso SAL	II Quotidiano magazine
17	Nova Siri	26/07/2004	4,445,645,0	631,600,0	64,0	4,0	16,00	0,13	Nova Siri SAL	Civil Protection
18	Valsinni	26/07/2004	4,448,638,8	624,031,9	86,6	6,0	14,43	0,13	Nova Siri SAL	Civil Protection
19	Melfi	20/09/2004	4,538,361,1	555,130,7	119,4	81,0	1,47	0,09	Melfi	Civil Protection
20	Montalbano Jonico	13/11/2004	4,460,236,3	634,365,1	67,2	33,0	2,04	0,23	Montalbano SAL	Civil Protection
21	Montescaglioso	13/11/2004	4,491,198,4	641,477,5	126,6	33,0	3,84	0,17	Montescaglioso SAL	Civil Protection

ID	Municipality	Date	UTM N	UTM E	H (mm)	D (h)	I (mm/h)	Antecedent soil saturation	Weather station	Sources
22	Pisticci	13/11/2004	4,470,264,9	643,085,0	94,6	34,0	2,78	0,25	Pisticci da Castelluccio SAL	Civil Protection
23	Tricarico	13/11/2004	4,498,622,9	582,231,5	46,2	31,0	1,49	0,32	Albano di Lucania PC	La Gazzetta del Mezzogiorno magazine
24	Tito	24/01/2005	4,493,939,0	556,898,4	40,0	17,0	2,35	0,72	Satriano di Lucania SAL	villasmunta.it
25	Nemoli	26/01/2005	4,435,028,2	568,166,9	76,2	47,0	1,62	0,90	Nemoli SAL	Civil Protection
26	San Fele	22/02/2005	4,518,910,4	546,690,8	77,0	80,0	0,96	0,97	San Fele PC	Civil Protection
27	Bella	23/02/2005	4,512,013,7	545,332,3	59,8	54,0	1,11	0,84	Bella Casalini	villasmunta.it
28	Castronuovo di Sant'Andrea	23/02/2005	4,449,641,7	600,766,2	26,0	30,0	0,87	0,64	Roccanova PC	Civil Protection
29	Picerno	23/02/2005	4,498,702,2	554,494,8	64,8	74,0	0,88	0,90	Balvano PC	villasmunta.it
30	Tito	24/02/2005	4,492,615,5	557,165,5	60,8	55,0	1,11	0,86	Satriano di Lucania SAL	villasmunta.it
31	Calvello	25/02/2005	4,478,560,2	573,724,4	55,4	56,0	0,99	0,94	Laurenzano PC	adnkronos.com
32	Rionero in Vulture	26/02/2005	4,527,555,0	558,531,6	14,2	6,0	2,37	0,72	Venosa SAL	villasmunta.it
33	Potenza	26/02/2005	4,498,664,2	567,834,7	64,4	98,0	0,66	0,87	Potenza PC	La Gazzetta del Mezzogiorno magazine
34	Potenza	26/02/2005	4,504,521,5	572,376,6	64,4	98,0	0,66	0,87	Potenza PC	Civil Protection
35	Sant'Arcangelo	26/02/2005	4,456,637,8	609,566,5	18,6	18,0	1,03	0,68	Roccanova PC	villasmunta.it
36	Gallicchio	27/02/2005	4,460,528,9	596,783,3	21,0	12,0	1,75	0,83	Guardia Perticara SAL	Civil Protection
37	Laurenzana	01/03/2005	4,482,440,8	578,878,4	86,6	148,0	0,59	0,93	Laurenzana SAL	Civil Protection
38	Pietrapertosa	02/03/2005	4,485,954,8	589,868,4	44,6	23,0	1,94	0,73	Campomaggiore SAL	Civil Protection
39	Nemoli	07/03/2005	4,438,133,6	570,116,6	45,8	69,0	0,66	0,93	Nemoli SAL	villasmunta.it
40	Barile	29/03/2005	4,533,072,0	557,767,6	25,8	29,0	0,89	0,64	Venosa SAL	Civil Protection

ID	Municipality	Date	UTM N	UTM E	H (mm)	D (h)	I (mm/h)	Antecedent soil saturation	Weather station	Sources
41	Accettura	07/06/2005	4,483,267,3	598,029,9	17,4	5,0	3,48	0,23	San Mauro Forte PC	Civil Protection
42	Terranova del Pollino	24/02/2006	4,426,066,5	610,796,4	79,6	78,0	1,02	0,72	Terranova del Pollino PC	Civil Protection
43	Bernalda	28/02/2006	4,474,473,5	648,787,6	44,0	11,0	4,00	0,72	Bernalda SAL	La Gazzetta del Mezzogiorno magazine
44	Grottole	28/02/2006	4,495,569,9	617,180,7	95,4	176,0	0,54	0,67	Grottole da Serre	La Gazzetta del Mezzogiorno magazine
45	Pisticci	28/02/2006	4,533,072,0	557,767,6	33,8	15,0	2,25	0,67	Torre Accio PC	La Gazzetta del Mezzogiorno magazine
46	Calvello	12/03/2006	4,479,726,4	575,486,3	50,6	29,0	1,74	0,92	Laurenzano PC	Basin Authority of Basilicata (AdB)
47	Montalbano Jonico	12/03/2006	4,472,188,9	640,718,0	37,8	56,0	0,68	0,64	Tursi SAL	Evalmet web site
48	Rionero in Vulture	13/03/2006	4,529,892,8	556,341,5	114,7	70,0	1,64	0,91	Melfi	Civil Protection
49	Venosa	13/03/2006	4,460,626,0	633,605,6	113,2	62,0	1,83	0,84	Venosa SAL	Civil Protection
50	Corleto Perticara	23/03/2006	4,475,673,7	588,722,5	43,6	43,0	1,01	0,83	Guardia Perticara SAL	AdB
51	Ripacandida	24/03/2006	4,529,316,2	560,944,2	30,5	58,0	0,53	0,77	Venosa SAL	Civil Protection
52	Picerno	27/03/2006	4,499,289,0	554,063,1	32,8	9,0	3,64	0,91	Balvano PC	Civil Protection
53	Trecchina	26/09/2006	4,430,915,4	567,168,1	133,0	72,0	1,85	0,63	Trecchina	Civil Protection
54	Rivello	23/10/2006	4,533,947,8	566,654,7	140,2	50,0	2,80	0,35	Nemoli SAL	IFFI Project ISPRA CNR IBAM
55	Maratea	19/12/2006	4,438,301,0	564,922,4	144,6	55,0	2,63	0,74	Maratea PC	infocilento
56	Maratea	04/04/2007	4,473,927,4	630,747,2	31,4	16,0	1,96	0,91	Maratea PC	infocilento
57	Tito	25/11/2008	4,429,377,2	561,850,9	77,4	120,0	0,65	0,43	Picerno PC	Civil Protection
58	Grassano	11/12/2008	4,492,851,3	557,094,7	79,4	16,0	4,96	0,45	Matera PC	La Gazzetta del Mezzogiorno
59	Calvello	05/01/2009	4,480,624,6	572,077,3	35,2	62,0	0,57	0,85	Laurenzano PC	Civil Protection

DOI: http://dx.doi.org/10.5772/intechopen.92730

ID	Municipality	Date	UTM N	UTM E	H (mm)	D (h)	I (mm/h)	Antecedent soil saturation	Weather station	Sources
60	Lagonegro	06/01/2009	4,502,711,0	621,802,2	109,8	65,0	1,69	0,88	Lagonegro PC	lucanianet
61	Grottole	13/01/2009	4,469,982,7	634,601,8	87,8	95,0	0,92	0,59	Grottole da Serre	II Quotidiano
62	Montescaglioso	13/01/2009	4,489,561,2	641,250,1	76,6	95,0	0,81	0,52	Montescaglioso SAL	La Gazzetta del Mezzogiorno
63	Pisticci	13/01/2009	4,487,771,2	600,527,1	105,2	114,0	0,92	0,62	Pisticci da Castelluccio SAL	Civil Protection
64	Pisticci	13/01/2009	4,495,207,5	617,127,3	104,6	113,0	0,93	0,62	Pisticci Scalo SAL	La Gazzetta del Mezzogiorno
65	Potenza	13/01/2009	4,499,028,2	571,863,4	17,6	20,0	0,88	0,94	Potenza PC	Civil Protection
66	Laurenzana	14/01/2009	4,477,838,6	584,070,9	38,8	35,0	1,11	0,95	Laurenzana SAL	Civil Protection
67	Acerenza	23/01/2009	4,515,988,0	578,994,5	32,0	28,0	1,14	0,58	Acerenza SAL	Civil Protection
68	Maratea	28/01/2009	4,441,638,3	566,470,6	295,2	185,0	1,60	0,93	Maratea PC	La Siritide website
69	Montalbano Jonico	06/03/2009	4,460,493,9	632,869,2	39,2	34,0	1,15	0,74	Montalbano SAL	Civil Protection
70	Pisticci	06/03/2009	4,429,791,9	561,758,0	40,6	34,0	1,19	0,85	Torre Accio PC	Evalmet web site
71	Tursi	06/03/2009	4,457,119,8	624,774,7	44,4	33,0	1,35	0,82	Tursi SAL	Civil Protection
72	Ripacandida	07/03/2009	4,471,046,5	639,954,0	93,5	69,0	1,36	0,62	Venosa SAL	palazzosangervasio.net
73	Gallicchio	20/03/2009	4,528,878,0	562,300,7	26,2	12,0	2,18	0,67	Aliano SAL	Civil Protection
74	Ripacandida	26/03/2009	4,528,798,8	561,165,2	22,8	36,0	0,63	0,79	Venosa SAL	Civil Protection
75	Tricarico	24/04/2009	4,524,102,2	555,355,7	75,0	135,0	0,56	0,90	Albano di Lucania PC	Civil Protection
76	San Martino D'Agri	28/04/2009	4,455,394,2	587,277,7	22,6	21,0	1,08	0,85	Sarconi SAL	La Gazzetta del Mezzogiorno
77	Vietri di Potenza	22/10/2009	4,500,015,6	596,374,9	11,6	9,0	1,29	0,68	Vietri	Quotidiano del sud, Metauronews
78	San Severino Lucano	18/12/2009	4,430,832,7	596,722,0	12,2	8,0	1,53	0,60	Viggianello SAL	Civil Protection
79	San Chirico Raparo	07/02/2010	4,448,987,6	591,663,1	29,0	5,0	5,80	0,83	Castelsaraceno PC	La Gazzetta del Mezzogiorno
80	Maratea	11/02/2010	4,427,557,0	562,256,5	56,6	76,0	0,74	0,96	Maratea PC	Civil Protection

ID	Municipality	Date	UTM N	UTM E	H (mm)	D (h)	I (mm/h)	Antecedent soil saturation	Weather station	Sources
81	Viggianello	13/02/2010	4,424,622,3	591,954,8	145,8	184,0	0,79	0,93	Viggianello SAL	Civil Protection
82	Latronico	20/02/2010	4,472,103,8	633,501,3	112,2	124,0	0,90	0,85	Castelsaraceno PC	II Quotidiano magazine
83	Tursi	11/10/2010	4,494,906,8	541,098,2	15,8	11,0	1,44	0,22	Tursi SI	Tursi tani.com
84	Ferrandina	02/11/2010	4,485,211,8	627,576,3	63,2	7,0	9,03	0,45	Ferrandina SAL	Civil Protection
85	Grottole	02/11/2010	4,493,102,4	615,427,5	115,2	7,0	16,46	0,46	Grottole da Serre	Civil Protection
86	Matera	02/11/2010	4,502,417,4	634,533,0	61,8	9,0	6,87	0,39	Matera PC	Civil Protection
87	Montescaglioso	02/11/2010	4,454,988,6	623,153,8	45,2	12,0	3,77	0,37	Montescaglioso SAL	II Quotidiano magazine
88	Pisticci	02/11/2010	4,491,414,3	544,109,2	73,4	10,0	7,34	0,44	Pisticci Scalo SAL	Civil Protection
89	Rivello	02/11/2010	4,437,163,0	564,328,3	38,4	10,0	3,84	0,82	Nemoli SAL	Civil Protection
90	Salandra	02/11/2010	4,486,232,1	612,353,1	108,0	6,0	18,00	0,32	San Mauro Forte PC	Civil Protection
91	Tursi	03/11/2010	4,456,064,4	625,339,9	62,3	11,0	5,66	0,39	Tursi SAL	Civil Protection
92	Melfi	10/11/2010	4,491,130,5	640,684,3	55,5	57,0	0,97	0,59	Melfi	II Quotidiano magazine
93	Potenza	11/11/2010	4,539,263,4	551,961,4	78,8	90,0	0,88	0,57	Potenza PC	La Gazzetta del Mezzogiorno magazine
94	Lauria	22/11/2010	4,504,815,2	572,162,3	169,8	140,0	1,21	0,87	Nemoli SAL	La Gazzetta del Mezzogiorno magazine
95	Muro Lucano	02/12/2010	4,440,667,8	573,120,0	135,4	250,0	0,54	0,79	Muro Lucano PC	La Gazzetta del Mezzogiorno magazine
96	Rivello	03/12/2010	4,512,300,7	536,385,6	248,2	271,0	0,92	0,90	Nemoli SAL	II Quotidiano del sud magazine
97	Castelluccio Inferiore	03/01/2011	4,428,643,1	583,828,8	50,8	41,0	1,24	0,86	Viggianello SAL	Civil Protection
98	Alianello	19/02/2011	4,437,455,5	564,407,9	36,8	10,0	3,68	0 57	Aliano SAL	ANAS (National Istitution for Highways)
99	Armento	19/02/2011	4,462,213,4	588,308,3	83,4	52,0	1,60	0,83	Guardia Perticara SAL	Civil Protection

ID	Municipality	Date	UTM N	UTM E	H (mm)	D (h)	I (mm/h)	Antecedent soil saturation	Weather station	Sources
100	Bernalda	19/02/2011	4,474,641,9	641,273,3	20,0	10,0	2,00	0,55	Bernalda SAL	Civil Protection
101	Montalbano Jonico	19/02/2011	4,461,782,0	633,367,9	28,4	17,0	1,67	0,48	Montalbano SAL	Civil Protection
102	Pisticci	19/02/2011	4,456,984,3	610,237,7	29,0	49,0	0,59	0,60	Pisticci Scalo SAL	Evalmet web site
103	Tursi	19/02/2011	4,457,443,9	626,287,3	30,8	24,0	1,28	0,39	Tursi SAL	Civil Protection
104	Valsinni	20/02/2011	4,447,737,2	622,905,0	22,8	16,0	1,43	0,59	Nova Siri SAL	Civil Protection
105	Cancellara	01/03/2011	4,509,339,7	577,969,2	42,8	14,0	3,06	0,93	San Nicola D'Avigliano PC	Civil Protection
106	Ferrandina	01/03/2011	4,485,235,5	624,000,0	99,0	19,0	5,21	0,70	Ferrandina SAL	Civil Protection
107	Matera	01/03/2011	4,501,551,8	634,766,1	103,0	23,0	4,48	0,58	Matera Nord SAL	Civil Protection
108	Teana	01/03/2011	4,442,436,7	598,621,8	64,2	21,0	3,06	0,90	Episcopia PC	Civil Protection
109	Bernalda	02/03/2011	4,474,369,9	642,163,4	102,4	21,0	4,88	0,58	Bernaida SAL	Civil Protection
110	Colobraro	02/03/2011	4,449,038,6	620,816,0	78,0	18,0	4,33	0,44	Tursi SI	metapontino.it
111	Grassano	02/03/2011	4,499,277,9	609,377,8	149,7	19,0	7,88	0,66	Grassano SAL	Evalmet web site
112	Irsina	02/03/2011	4,456,172,0	625,292,0	89,4	21,0	4,26	0,19	Santa Marla d'Irsi SAL	pisticci.com
113	Montalbano Jonico	02/03/2011	4,471,789,5	633,568,6	153,6	24,0	6,40	0,52	Montalbano SAL	Evalmet web site
114	Pisticci	02/03/2011	4,449,746,0	630,633,3	81,4	18,0	4,52	0,65	Pisticci Scalo SAL	II Quotidiano, Evalmet web site
115	Rotondella	02/03/2011	4,518,058,7	611,730,4	126,6	17,0	7,45	0,65	Nova Siri SAL	Civil Protection
116	Tricarico	02/03/2011	4,496,864,3	597,934,1	31,4	15,0	2,09	0,82	Albano di Lucania PC	vigilfuoco.it
117	Tursi	02/03/2011	4,460,225,2	632,086,0	152,0	18,0	8,44	0,44	Tursi SAL	Evalmet web site
118	Valsinni	02/03/2011	4,472,089,5	632,234,9	129,2	18,0	7,18	0,65	Nova Siri SAL	Evalmet web site
119	Laurenzana	03/03/2011	4,478,608,5	582,713,6	81,6	63,0	1,30	0,90	Laurenzana SAL	Evalmet web site
120	Lauria	05/03/2011	4,432,403,8	572,113,3	97,8	114,0	0,86	0,84	Nemoli SAL	Civil Protection

ID	Municipality	Date	UTM N	UTM E	H (mm)	D (h)	I (mm/h)	Antecedent soil saturation	Weather station	Sources
121	Grottole	05/05/2011	4,447,270,2	625,680,9	62,8	102,0	0,62	0,94	Grottole da Castellano	provincia di Matera
122	Laurenzana (riatt.)	05/05/2011	4,481,944,5	581,225,0	149,8	210,0	0,71	0,78	Laurenzana SAL	ANAS
123	Bella	07/10/2011	4,512,353,3	546,253,8	94,2	5,0	18,84	0,17	Bella Casalini	Civil Protection
124	Muro Lucano	08/10/2011	4,511,194,6	540,805,5	127,6	6,0	21,27	0,07	Muro Lucano PC	La Gazzetta del Mezzogiorno magazine
125	Matera	06/11/2011	4,502,097,8	635,438,5	34,2	2,0	17,10	0,10	Matera PC	Civil Protection
126	San Fele	06/12/2011	4,515,972,5	551,039,6	23,8	30,0	0,79	0,44	San Fele PC	Civil Protection
127	Latronico	15/11/2011	4,437,998,4	586,199,8	65,0	44,0	1,48	0,89	Episcopia PC	Civil Protection
128	Stigliano	25/12/2011	4,473,739,4	605,248,1	46,6	8,0	5,83	0,13	Stigliano SAL	Civil Protection
129	Lauria	06/01/2012	4,431,454,0	572,485,4	41,6	46,0	0,90	0,82	Nemoli SAL	Civil Protection
130	Rivello (2)	20/01/2012	4,436,157,6	558,533,0	16,6	10,0	1,66	0,80	Nemoli SAL	Civil Protection
131	Savoia di Lucania	04/02/2012	4,490,321,5	547,596,1	57,2	79,0	0,72	0,37	Vietri	Civil Protection
132	Craco	08/02/2012	4,470,541,6	622,549,9	37,4	24,0	1,56	0,38	Craco PC	Civil Protection
133	Rapone	10/02/2012	4,521,844,9	542,278,7	20,1	17,0	1,18	0,53	San Fele PC	Civil Protection
134	Montemurro	11/02/2012	4,461,336,3	583,827,5	118,6	254,0	0,47	0,54	Grumento Nova	Civil Protection
135	Avigliano	12/02/2012	4,509,076,7	560,722,3	29,8	39,0	0,76	0,61	Avigliano PC	Civil Protection
136	Bernalda	23/02/2012	4,475,260,0	643,993,1	52,4	42,0	1,25	0,48	Bernalda SAL	Civil Protection
137	Castronuovo di Sant'Andrea	23/02/2012	4,448,673,1	604,075,0	95,4	50,0	1,91	0,48	Roccanova PC	basilicatanotizie.net
138	Chlaromonte	23/02/2012	4,499,219,0	647,309,0	100,4	45,0	2,23	0,51	Noepoli PC	basilicatanotizie.net
139	Montalbano Jonico	23/02/2012	4,461,399,7	632,113,0	46,8	43,0	1,09	0,51	Montalbano SAL	Civil Protection
140	Tursi	23/02/2012	4,456,338,6	624,326,0	53,8	43,0	1,25	0,63	Tursi SAL	Civil Protection

ID	Municipality	Date	UTM N	UTM E	H (mm)	D (h)	I (mm/h)	Antecedent soil saturation	Weather station	Sources
141	Pietrapertosa	24/02/2012	4,440,620,5	607,006,8	39,2	39,0	1,01	0,61	Campomaggiore SAL	Civil Protection
142	Vietri di Potenza	08/03/2012	4,495,336,2	543,776,2	10,0	7,0	1,43	0,50	Vietri	Quotidiano del sud, Metauronews
143	Avigliano	09/03/2012	4,507,355,3	561,591,1	31,4	18,0	1,74	0,71	Avigliano PC	Civil Protection
144	Rivello	14/04/2012	4,436,142,6	554,613,5	118,0	30,0	3,93	0,71	Nemoli SAL	Civil Protection
145	Rapone	18/04/2012	45,873,9	542,457,0	50,6	KO	0,56	0,76	San Fele PC	Civil Protection
146	Avigliano	20/04/2012	4,509,561,9	561,009,0	20,4	33,0	0,62	0,76	Avigliano PC	Civil Protection
147	Rivello	21/04/2012	4,435,386,6	565,318,9	256,8	192,0	1,34	0,71	Nemoli SAL	Civil Protection
148	Lauria	06/06/2012	4,432,952,2	570,399,6	58,2	11,0	5,29	0,76	Nemoli SAL	Civil Protection
149	Teana	23/06/2012	4,442,695,5	598,168,9	29,6	3,0	9,87	0,30	Episcopia PC	Civil Protection
150	Lavello	01/09/2012	4,545,949,2	568,245,9	20,8	4,0	5,20	0,02	Lavello SAL	Civil Protection
151	Venosa	02/09/2012	4,460,329,4	633,421,4	141,0	32,0	4,41	0,08	Venosa SAL	Civil Protection
152	Castelluccio Inferiore (3)	03/10/2012	4,427,492,7	587,471,1	9,0	2,0	4,50	0,12	Viggianello SAL	Civil Protection
153	Rotonda	29/10/2012	4,423,198,3	588,915,1	111,2	28,0	3,97	0,27	Rotonda SAL	Civil Protection
154	Campomaggiore	20/11/2012	4,491,247,0	590,935,1	66,8	106,0	0,63	0,24	Campomaggiore SAL	Civil Protection
155	Pisticci	20/11/2012	4,533,376,1	575,381,2	73,8	67,0	1,10	0,54	Pisticci Scalo SAL	pisticci.com
156	Roccanova	20/11/2012	4,453,378,3	603,847,3	96,2	90,0	1,07	0,30	Roccanova PC	Civil Protection
157	Vietri di Potenza	20/11/2012	4,494,560,8	543,083,9	27,0	29,0	0,93	0,44	Vietri	Quotidiano del sud, Metauronews
158	Lauria	04/12/2012	4,472,662,9	632,178,6	220,0	152,0	1,45	0,80	Nemoli SAL	regione.basilicata.it
159	Barile	08/12/2012	4,532,258,7	556,671,8	26,1	12,0	2,18	0,46	Venosa SAL	Civil Protection
160	San Severino Lucano	17/01/2013	4,430,496,6	596,772,9	175,6	177,0	0,99	0,85	Viggianello SAL	Civil Protection
161	Lauria	18/01/2013	4,434,754,0	571,329,9	268,0	180,8	1,49	0,79	Nemoli SAL	Civil Protection
162	Savoia di Lucania	18/01/2013	4,491,704,0	544,971,0	87,6	93,0	0,94	0,68	Vietri	Civil Protection

ID	Municipality	Date	UTM N	UTM E	H (mm)	D (h)	I (mm/h)	Antecedent soil saturation	Weather station	Sources
163	Vietri di Potenza	18/01/2013	4,432,069,0	573,009,3	87,6	93,0	0,94	0,73	Vietri	Quotidiano del sud, Metauronews
164	Sant'Angelo le Fratte	19/01/2013	4,494,125,0	541,429,1	99,6	101,0	0,99	0,63	Tito PC	Fonti Cronachistiche
165	Lagonegro	21/01/2013	4,488,383,1	547,262,1	266,8	244,0	1,09	0,80	Lagonegro PC	La Siritide
166	Episcopia	25/01/2013	4,445,478,1	562,707,3	366,6	362,0	1,01	0,87	Episcopia PC	basilicatanotizie.net
167	San Severino Lucano	03/02/2013	4,429,936,5	596,946,0	28,2	26,0	1,08	0,90	Viggianello SAL	Civil Protection
168	Armento	13/02/2013	4,462,461,3	590,367,6	12,0	10,0	1,20	0,73	Guardia Perticara SAL	Civil Protection
169	Balvano	13/02/2013	4,500,024,3	543,479,7	34,4	31,0	1,11	0,74	Balvano PC	Civil Protection
170	Avigliano	24/02/2013	4,508,495,2	559,778,8	18,4	27,0	0,68	0,95	Avigliano PC	Civil Protection
171	Vietri di Potenza	14/03/2013	4,437,619,1	594,262,1	44,8	49,0	0,91	0,89	Vietri	Quotidiano del sud, Metauronews
172	Castelluccio Inferiore	15/03/2013	4,428,429,2	584,348,2	136,0	214,0	0,64	0,84	Viggianello SAL	Civil Protection
173	Lauria	21/03/2013	4,495,266,6	544,987,6	339,2	353,0	0,96	0,82	Nemoli SAL	youtube
174	Castelluccio Inferiore	10/04/2013	4,428,714,4	584,450,1	12,2	3,0	4,07	0,83	Viggianello SAL	Civil Protection
175	Muro Lucano	10/07/2013	4,510,677,3	541,344,5	59,2	53,0	1,12	0,36	Muro Lucano PC	Civil Protection
176	Vietri di Potenza	21/07/2013	4,493,924,6	541,919,0	13,8	4,0	3,45	0,32	Vietri	Quotidiano del sud, Metauronews
177	Accettura	21/08/2013	4,428,008,9	574,521,3	27,4	25,0	1,10	0,19	Campomaggiore SAL	accettura online
178	Atella - Filiano	21/08/2013	4,517,964,3	559,520,9	43,6	5,0	8,72	0,21	Atella PC	Civil Protection
179	Bernalda	07/10/2013	4,474,073,6	643,387,8	190,4	56,0	3,40	0,08	Bernalda SAL	Quotidiano del sud, Evilmet
180	Montalbano Jonico	07/10/2013	4,461,773,7	634,506,4	84,2	2,90	2,90	0,10	Montalbano SAL	Civil Protection
181	Montescaglioso	07/10/2013	4,490,286,3	641,544,0	135,2	35,0	3,86	0,07	Montescaglioso SAL	Civil Protection
182	Pisticci	07/10/2013	4,487,593,1	594,710,1	105,6	57,0	1,85	0,11	Torre Accio PC	Evalmet web site
183	Lauria	11/10/2013	4,470,914,4	645,919,6	43,4	51,0	0,85	0,39	Nemoli SAL	infopinione.it
184	Bernalda	16/11/2013	4,472,586,8	647,193,6	148,4	136,0	1,09	0 37	Metaponto	Gazzettino.it

ID	Municipality	Date	UTM N	UTM E	H (mm)	D (h)	I (mm/h)	Antecedent soil saturation	Weather station	Sources
185	Chiaromonte	16/11/2013	4,433,677,5	571,261,9	238,2	132,0	1,80	0,22	Noepoli PC	La Siritide
186	Pisticci	16/11/2013	4,472,993,9	631,724,0	157,0	135,0	1,16	0,24	Pisticci Scalo SAL	Civil Protection
187	San Fele	22/11/2013	4,518,881,1	545,411,7	109,8	82,0	1,34	0,42	San Fele PC	Civil Protection
188	Guardia Perticara (3)	24/11/2013	4,468,838,7	592,161,9	103,6	115,9	0,90	0,61	Guardia Perticara SAL	Civil Protection
189	San Severino Lucano	24/11/2013	4,429,132,5	594,345,3	133,8	100,0	1,34	0,58	Viggianello SAL	lapretoria.it
190	Laurenzana	26/11/2013	4,440,683,6	606,559,5	89,0	94,0	0,95	0,61	Laurenzana SAL	Civil Protection
191	Valsinni	26/11/2013	4,446,786,7	622,424,8	193,0	115,0	1,68	0,25	Nova Siri SAL	Civil Protection
192	Miglionico	30/11/2013	4,492,223,8	627,675,2	20,0	8,0	2,50	0,48	Ferrandina SAL	Civil Protection
193	Chiaromonte	01/12/2013	4,479,622,4	582,489,7	65,0	25,0	2,60	0,85	Episcopia PC	Fonti Cronachistiche
194	Craco (7)	01/12/2013	4,470,341,9	623,096,0	156,4	29,0	5,39	0,61	Craco PC	Civil Protection
195	Gallicchio	01/12/2013	4,489,127,2	591,325,4	95,6	20,0	4,78	0,54	Aliano SAL	Civil Protection
196	Ginestra	01/12/2013	4,531,394,8	562,130,8	89,5	23,0	3,89	0,32	Venosa SAL	Civil Protection
197	Pomarico (4)	01/12/2013	4,486,804,1	631,164,0	128,6	26,0	4,95	0,51	Ferrandina SAL	Civil Protection
198	Potenza	01/12/2013	4,496,760,3	571,166,4	63,2	25,0	2,53	0,60	Potenza PC	Civil Protection
199	Savoia di Lucania	01/12/2013	4,490,996,4	546,572,9	125,6	51,0	2,46	0,63	Vietri	Civil Protection
200	Tricarico	01/12/2013	4,497,852,6	597,122,7	114,8	32,0	3,59	0,55	Albano di Lucania PC	Civil Protection
201	Armento	02/12/2013	4,449,866,2	621,395,1	120,0	57,0	2,11	0,81	Guardia Perticara SAL	Fonti Cronachistiche
202	Bernalda	02/12/2013	4,473,912,5	643,362,6	221,3	49,0	4,52	0,74	Bernalda SAL	Civil Protection
203	Cirigliano	02/12/2013	4,472,564,9	599,264,5	162,0	56,0	2,89	0,59	San Mauro Forte PC	Civil Protection
204	Colobraro	02/12/2013	4,461,913,3	596,895,9	140,7	26,0	5,41	0,68	Sinni a Valsinni SI	emmenews
205	Garaguso/Grassano	02/12/2013	4,545,101,3	565,137,4	150,0	52,0	2,88	0,47	Grassano SAL	La Gazzetta del Mezz
206	Grottole(3)	02/12/2013	4,494,492,3	616,732,9	186,8	57,0	3,28	0,48	Grottole da Serre	Civil Protection

ID	Municipality	Date	UTM N	UTM E	H (mm)	D (h)	I (mm/h)	Antecedent soil saturation	Weather station	Sources
207	Guardia Perticara	02/12/2013	4,494,394,2	605,039,5	120,0	57,0	2,11	0,81	Guardia Perticara SAL	Civil Protection
208	Lavello	02/12/2013	4,459,823,2	599,087,0	130,0	57,0	2,28	0,30	Lavello SAL	Fonti Cronachistiche
209	Missanello	02/12/2013	4,462,814,9	590,380,8	138,6	55,0	2,52	0,54	Aliano SAL	Fonti Cronachistiche
210	Montalbano Jonico	02/12/2013	4,457,687,7	626,816,0	174,8	28,0	6,24	0,66	Montalbano SAL	Civil Protection
211	Pietrapertosa	02/12/2013	4,484,995,6	589,793,9	168,8	54,0	3v13	0,57	Campomaggiore SAL	Civil Protection
212	Pisticci	02/12/2013	4,466,906,3	594,785,7	167,2	33,0	5,07	0,63	Pisticci Scalo SAL	Civil Protection
213	Rivello	02/12/2013	4,436,644,0	564,876,7	42,2	38,0	1,11	088	Nemoli SAL	Civil Protection
214	Senise	02/12/2013	4,445,683,3	609,339,4	83,0	37,0	2,24	0,81	Noepoli PC	Civil Protection
215	Accettura	03/12/2013	4,488,048,3	593,820,9	211,4	58,0	3,64	0,57	Campomaggiore SAL	La Gazzetta del Mezzogiorno
216	Accettura, Salandra	03/12/2013	4,481,543,0	598,406,7	197,8	63,0	3,14	0,59	San Mauro Forte PC	La Gazzetta del Mezzogiorno
217	Ferrandina	03/12/2013	4,436,089,0	624,424,9	161,0	59,0	2,73	0,51	Ferrandina SAL	Civil Protection
218	Ginestra	03/12/2013	4,531,331,9	561,335,0	128,5	56,0	2,29	0,32	Venosa SAL	Civil Protection
219	Montescaglioso	03/12/2013	4,489,336,0	639,955,0	224,2	56,0	4,00	0,54	Montescaglioso SAL	Pellicani et al., 2016
220	Trivigno (4)	03/12/2013	4,490,716,4	583,663,1	196,0	88,0	2,23	0,55	Albano di Lucania PC	Civil Protection
221	Tursi	03/12/2013	4,472,751,0	632,461,7	138,6	66,0	2,10	0,58	Tursi SAL	Evalmet web site
222	Viggianello	04/12/2013	4,427,183,0	589,852,6	74,8	61,0	1,23	0,35	Viggianello SAL	Civil Protection
223	Sarconi	26/12/2013	4,457,355,0	632,845,6	26,0	11,0	2,36	0,85	Sarconi SAL	Civil Protection
224	Aliano	21/01/2014	4,463,968,1	603,952,3	88,4	52,0	1,70	0,83	Roccanova PC	Fonti Cronachistiche
225	Calvera	21/01/2014	4,454,657,0	574,796,0	193,6	69,9	2,81	0,86	Episcopia PC	La Siritide
226	Castronuovo di Sant'Andrea	21/01/2014	4,450,030,0	600,486,2	87,4	60,0	1,46	0,83	Roccanova PC	Civil Protection
227	Rivello	21/01/2014	4,436,820,3	566,903,7	114,8	41,0	2,80	0,86	Nemoli SAL	Civil Protection

ID	Municipality	Date	UTM N	UTM E	H (mm)	D (h)	I (mm/h)	Antecedent soil saturation	Weather station	Sources
228	Senise	21/01/2014	4,444,739,2	610,507,1	52,0	50,0	1,04	0,84	Senise SAL	Civil Protection
229	Latronico	22/01/2014	4,436,079,5	583,979,5	194,9	70,0	2,77	0,86	Episcopia PC	La Siritide website
230	Lauria	22/01/2014	4,444,641,2	597,413,0	267,2	86,0	3,11	0,86	Nemoli SAL	Fonti Cronachistiche
231	Guardia Perticara	24/01/2014	4,469,389,1	594,580,2	103,6	128,0	0,81	0,85	Guardia Perticara SAL	Civil Protection
232	Maratea	28/01/2014	4,427,583,9	561,128,4	234,6	218,0	1,08	0,93	Maratea PC	Civil Protection
233	Francavilla in Sinni	30/01/2014	4,433,285,2	569,276,0	247,8	238,0	1,04	0,86	Episcopia PC	regione.basilicata.it
234	Guardia Perticara	02/02/2014	4,467,955,0	593,670,0	46,4	46,0	1,01	0,95	Guardia Perticara SAL	Coldiretti webpage
235	Potenza	02/02/2014	4,496,094,5	568,378,3	12,4	7,0	1,77	0,91	Potenza PC	Civil Protection
236	Tursi	02/02/2014	4,436,149,4	600,183,4	30,6	36,0	0,85	0,86	Tursi SAL	oltrefreepress.com
237	Accettura	03/02/2014	4,482,603,0	600,241,7	105,6	77,0	1,37	0,79	San Mauro Forte PC	accettura online
238	Castelluccio Inferiore	03/02/2014	4,428,421,9	583,511,5	272,6	381,0	0,72	0,78	Viggianello SAL	castelluccioinferiore.comune.news
239	Gallicchio	03/02/2014	4,460,176,1	597,304,5	88,0	67,0	1,31	0,92	Roccanova PC	Civil Protection
240	Latronico	03/02/2014	4,438,541,9	586,536,2	47,6	68,0	0,70	0,94	Episcopia PC	Civil Protection
241	Pisticci	03/02/2014	4,458,130,1	631,793,0	62,4	68,0	0,92	0,87	Craco PC	Fonti Cronachistiche magazine
242	San Giorgio Lucano	03/02/2014	4,441,730,4	618,593,4	60,2	74,0	0,81	0,94	San Giorgio Lucano SAL	Civil Protection
243	Aliano	04/02/2014	4,462,662,3	608,499,5	94,6	81,0	1,17	0,92	Roccanova PC	Civil Protection
244	Gorgoglione	04/02/2014	4,471,971,5	597,509,0	86,2	74,0	1,16	0,95	Guardia Perticara SAL	basilicata24.it
245	Guardia Perticara (2)	04/02/2014	4,468,759,4	593,490,1	91,8	85,0	1,08	0,95	Guardia Perticara SAL	Civil Protection
246	Missanello(2)	04/02/2014	4,459,645,8	598,794,7	96,6	81,0	1,19	0,79	Aliano SAL	Civil Protection
247	Noepoli	04/02/2014	4,473,633,7	637,169,5	101,6	74,0	1,37	0,90	Noepoli PC	La Siritide
248	Sant'Arcangelo (3)	04/02/2014	4,454,767,9	612,151,0	94,6	81,0	1,17	0,92	Roccanova PC	Civil Protection
249	Cirigliano	05/02/2014	4,476,117,9	599,525,8	124,8	90,0	1,39	0,79	San Mauro Forte PC	sassiland

ID	Municipality	Date	UTM N	UTM E	H (mm)	D (h)	I (mm/h)	Antecedent soil saturation	Weather station	Sources
250	Sarconi	09/02/2014	4,473,389,5	604,347,8	22,4	30,0	0,75	0,93	Sarconi SAL	Civil Protection
251	Stigliano	09/02/2014	4,438,261,8	613,374,0	105,6	98,0	1,08	0,78	Stigliano SAL	Civil Protection
252	Armento	25/03/2014	4,461,174,2	591,120,8	17,4	14,0	1,24	0,81	Guardia Perticara SAL	Civil Protection
253	Guardia Perticara	25/03/2014	4,455,948,6	575,716,1	30,8	16,0	1,93	0,81	Guardia Perticara SAL	Civil Protection
254	Montemurro	27/03/2014	4,462,003,7	585,867,7	18,2	20,0	0,91	0,89	Grumento NOva	Civil Protection
255	Stigliano	27/03/2014	4,473,506,8	604,550,1	12,0	4,0	3,00	0,68	Stigliano SAL	regione.basilicata.it
256	San Severino Lucano	06/04/2014	4,429,371,0	597,800,0	68,0	49,0	1,39	0,89	Viggianello SAL	sanseverinolucano.com
257	Rionero in Vulture	12/04/2014	4,530,903,2	556,997,1	29,1	16,0	1,82	0,74	Melfi	Civil Protection
258	Chiaromonte	16/04/2014	4,470,704,8	595,848,0	21,0	13,0	1,62	0,87	Noepoli PC	Civil Protection
259	Montemurro	17/04/2014	4,460,278,1	587,345,4	13,2	7,0	1,89	0,83	Grumento Nova	Civil Protection
260	San Severino Lucano	30/04/2014	4,439,756,0	605,195,5	37,4	47,0	0,80	0,87	Viggianello SAL	Civil Protection
261	Calvello	24/07/2014	4,427,437,3	600,212,2	11,0	1,0	11,00	0,31	Laurenzana SAL	Civil Protection
262	Rivello	08/11/2014	4,435,948,7	554,641,6	53,4	19,0	2,81	0,30	Nemoli SAL	Civil Protection
263	Rionero in Vulture (2)	31/12/2014	4,530,984,6	556,617,6	14,0	6,0	2,33	0,43	Melfi	Civil Protection
264	Brienza	30/01/2015	4,481,683,5	553,437,4	89,8	24,0	3,74	0,75	Brienza PC	Civil Protection
265	Chiaromonte	30/01/2015	4,442,083,4	603,091,6	36,8	53,0	0,69	0,53	Senise SAL	Civil Protection
266	Senise	30/01/2015	4,444,603,9	607,201,7	36,8	53,0	0,69	0,58	Senise SAL	La Siritide website
267	Castelsaraceno	31/01/2015	4,435,933,3	568,128,3	157,6	36,0	4,38	0,81	Castelsaraceno SI	Civil Protection
268	Lagonegro	31/01/2015	4,429,110,9	594,311,5	284,9	53,0	5,36	0,87	Lagonegro PC	Civil Protection
269	Latronico	31/01/2015	4,435,217,2	590,205,7	164,9	61,0	2,69	0,78	Episcopia PC	Civil Protection
270	Lauria	31/01/2015	4,448,327,7	565,843,8	263,8	43,0	6,13	0,88	Nemoli SAL	meteoweb.eu
271	Nemoli	31/01/2015	4,427,708,0	574,027,3	263,8	43,0	6,13	0,92	Nemoli SAL	Civil Protection

ID	Municipality	Date	UTM N	UTM E	H (mm)	D (h)	I (mm/h)	Antecedent soil saturation	Weather station	Sources
272	San Costantino Albanese	31/01/2015	4,432,781,2	612,691,0	61,6	80,0	0,77	0,49	Terranova del Pollino PC	Civil Protection
273	San Martino D'Agri	31/01/2015	4,454,871,1	589,437,1	73,2	34,0	2,15	0,74	Roccanova PC	meteoweb.eu
274	Viggianello	31/01/2015	4,447,312,7	586,781,1	164,0	61,0	2,69	0,81	Episcopia PC	meteoweb.eu
275	Montemurro	04/02/2015	4,461,149,4	584,066,8	144,6	132,0	1,10	0,54	Grumento Nova	Fonti Cronachistiche magazine
276	San Severino Lucano	06/02/2015	4,426,488,0	595,925,0	315,8	240,0	1,32	0,82	Viggianello SAL	regione.basilicata.it
277	Trecchina	08/02/2015	4,428,416,7	568,127,6	344,0	248,0	1,39	0,95	Castrocucco	Fonti Cronachistiche
278	Lauria	06/03/2015	4,434,254,5	570,436,9	99,2	82,0	1,21	0,86	Nemoli SAL	La Siritide website
279	Tricarico	06/03/2015	4,497,116,3	597,635,0	40,4	40,0	1,01	0,66	Albano di Lucania PC	tricarico news
280	Vietri di Potenza	12/03/2015	4,494,061,7	543,185,2	9,6	5,0	1,92	0,84	Vietri	Quotidiano del sud magazine
281	Terranova del Pollino	16/03/2015	4,425,194,1	607,145,4	85,2	120,0	0,71	0,84	Terranova del Pollino PC	basilicatanotizie.net
282	Salandra	18/03/2015	4485066v9	613,268,6	13,8	20,0	0,69	0,72	San Mauro Forte PC	Civil Protection
283	Montemurro	25/03/2015	4,461,303,7	585,075,4	31,4	62,0	0,51	0,88	Grumento Nova	rainews
284	Colobraro	26/03/2015	4,449,328,3	620,218,2	30,2	9,0	3,36	0,76	Sinni a Valsinni SI	Civil Protection
285	Ferrandina	27/03/2015	4,472,778,5	627,699,5	99,2	72,0	1,38	0,80	Ferrandina SAL	youreporter
286	Calvera	28/03/2015	4,445,756,6	595,749,2	80,6	128,0	0,63	0,90	Episcopia PC	Civil Protection
287	Castronuovo di Sant'Andrea	28/03/2015	4,449,432,7	600,769,2	83,4	80,0	1,04	0,91	Roccanova PC	Civil Protection
288	Chiaromonte	28/03/2015	4,441,944,5	603,702,8	71,6	78,0	0,92	0,73	Noepoli PC	Civil Protection
289	Colobraro	28/03/2015	4,451,661,8	625,343,5	78,5	77,0	1,02	0,76	Sinni a Valsinni SI	La Gazzetta del Mezzogiorno magazine
290	Grottole	28/03/2015	4,495,416,1	614,736,1	107,2	123,0	0,87	0,64	Grottole da Serre	Civil Protection

ID	Municipality	Date	UTM N	UTM E	H (mm)	D (h)	I (mm/h)	Antecedent soil saturation	Weather station	Sources
291	Terranova del Pollino	28/03/2015	4,424,299,8	606,589,5	142,4	129,0	1,10	0,86	Terranova del Pollino PC	Civil Protection
292	Anzi	06/04/2015	4,484,904,3	579,311,8	31,6	66,0	0,48	0,89	Laurenzana SAL	Civil Protection
293	Sasso di Castalda	12/06/2015	4,482,734,2	556,250,0	178,8	76,0	2,35	0,59	Brienza PC	Civil Protection
294	Grassano	11/08/2015	4,498,091,9	608,264,6	106,2	28,0	3,79	0,15	Grassano SAL	Civil Protection
295	Grottole	11/08/2015	4,497,727,3	612,882,2	86,0	35,0	2,46	0,15	Grottole da Serre	meteoweb.eu
296	Matera	11/08/2015	4,500,866,1	633,780,7	54,3	51,0	1,06	0,11	Matera PC	Civil Protection
297	Melfi	11/08/2015	4,538,822,6	554,831,5	11,8	2,0	5,91	0,04	Melfi	Civil Protection
298	Salandra	11/08/2015	4,487,314,1	610,376,5	21,8	26,0	0,84	0,22	San Mauro Forte PC	Civil Protection
299	Venosa	20/10/2015	4,537,376,0	569,867,0	22,8	15,0	1,52	0,24	Venosa SAL	Civil Protection
300	Chiaromonte	31/10/2015	4,442,678,8	603,349,5	108,0	51,0	2,12	0,30	Senise SAL	Civil Protection
301	Gallicchio	06/01/2016	4,460,397,1	596,927,3	10,6	3,0	3,53	0,39	Aliano SAL	La Siritide website
302	Rotonda	13/02/2016	4,423,399,8	588,125,9	117,0	91,0	1,29	0,77	Rotonda SAL	Civil Protection
303	Picerno	15/02/2016	4,500,954,0	549,546,0	93,2	103,0	0,90	0,51	Balvano PC	Civil Protection
304	Venosa	18/02/2016	4,535,187,4	568,991,4	10,7	8,0	1,34	0,38	Venosa SAL	Civil Protection
305	Castronuovo di Sant'Andrea	13/03/2016	4,449,163,2	601,141,5	120,6	50,0	2,41	0,37	Roccanova PC	Civil Protection
306	Terranova del Pollino	14/03/2016	4,425,609,1	607,819,7	63,4	46,0	1,38	0,79	Terranova del Pollino PC	Civil Protection
307	Ferrandina	16/03/2016	4,472,610,0	638,594,0	101,0	150,0	0,67	0,40	Ferrandina SAL	youreporter
308	Sant'Arcangelo (2 eventi)	16/03/2016	4,452,238,1	611,364,0	145,6	48,0	3,03	0,38	Aliano SAL	Civil Protection
309	Colobraro	17/03/2016	4,449,758,6	620,171,4	321,4	144,0	2,23	0,63	Sinni a Valsinni SI	Civil Protection

ID	Municipality	Date	UTM N	UTM E	H (mm)	D (h)	I (mm/h)	Antecedent soil saturation	Weather station	Sources
310	Pisticci	17/03/2016	4,482,354,3	626,252,6	72,1	26,0	2,77	0,57	Torre Accio PC	youreporter
311	Stigliano	17/03/2015	4,472,596,0	604,150,0	238,2	142,0	1,68	0,36	Stigliano SAL	Civil Protection
312	Salandra	18/03/2016	4,485,538,3	613,656,0	156,6	192,0	0,82	0,41	Salandra SI	Civil Protection
313	Pisticci	25/03/2016	4,471,998,1	633,051,0	47,0	18,0	2,61	0,73	Torre Accio PC	Civil Protection
314	Sant'Angelo le Fratte	26/03/2016	4,487,108,9	547,444,0	35,8	33,0	1,08	0,68	Tito PC	Civil Protection
315	Miglionico	26/07/2016	4,492,210,7	626,535,0	13,6	1,0	13,60	0,17	Ferrandina PC	Quotidiano della Basilicata magazine
316	Lavello	11/09/2016	4,545,245,7	564,838,8	31,8	19,0	1,67	0,21	Lavello SI	vulturenews
317	Genzano	21/09/2016	4,521,410,4	590,384,9	19,4	6,0	3,23	0,33	Genzano SAL	Civil Protection
318	Maratea	10/10/2016	4,426,721,0	560,774,8	113,4	70,0	1,62	0,27	Maratea PC	Fonti Cronachistiche
319	Colobraro	23/01/2017	4,447,990,5	622,489,0	127,2	68,0	1,87	0,57	Sinni a Valsinni SI	Civil Protection
320	Campomaggiore	25/01/2017	4,490,127,7	591,486,1	67,2	67,0	1,00	0,67	Laurenzano PC	Civil Protection
321	Senise	25/01/2017	4,445,316,4	612,080,5	113,0	91,0	1,24	0,52	Noepoli PC	trmtv.it
322	Avigliano	08/03/2017	4,510,867,0	560,723,4	53,6	31,0	1,73	0,86	Avigliano PC	basilicata24.com
323	Montemilone	15/07/2017	4,542,486,2	581,337,0	27,4	2,0	13,70	0,05	Montemilone PC	vulture news
324	Viggianello	05/03/2018	4,426,435,3	589,650,4	123,0	197,0	0,62	0,87	Rotonda	Civil Protection
325	Rivello	07/03/2018	4,436,820,1	565,132,1	208,0	360,0	0,58	0,88	Lagonegro PC	Civil Protection
326	Laurenzana	28/03/2018	4,479,367,5	582,543,4	108,2	192,0	0,56	0,83	Laurenzana	Civil Protection

Table 1.
Rainfall-triggered landslides in Basilicata during last 20 years.

Author details

Maurizio Lazzari[1], Marco Piccarreta[2], Ram L. Ray[3] and Salvatore Manfreda[4*]

1 CNR-ISPC, Potenza (PZ), Tito Scalo, Italy

2 Interuniversity Consortium for Hydrology (CINID), Potenza, Italy

3 College of Agriculture and Human Sciences, Prairie View A&M University, Prairie View, Texas, USA

4 Department of Civil, Architectural and Environmental Engineering, University of Naples Federico II, Naples, Italy

*Address all correspondence to: salvatore.manfreda@unina.it

References

[1] Hong Y, Adler R, Huffman G. Evaluation of the potential of NASA multi-satellite precipitation analysis in global landslide hazard assessment. Geophysical Research Letters. 2006;**33**: L22402. DOI: 10.1029/2006GL028010

[2] Marc O, Stumpf A, Malet JP, Gosset M, Uchida T, Chiang SH. Initial insights from a global database of rainfall-induced landslide inventories: The weak influence of slope and strong influence of total storm rainfall. Earth Surface Dynamics. 2018;**6**:903-922

[3] Guzzetti F, Peruccacci S, Rossi M, Stark CP. The rainfall intensity-duration control of shallow landslides and debris flows: An update. Landslides. 2008;**5**:3-17

[4] Lazzari M, Piccarreta M, Capolongo D. Landslide triggering and local rainfall thresholds in Bradanic Foredeep, Basilicata region (southern Italy). Landslide Science and Practice. Vol. 2. Early Warning, Instrumentation and Modeling. Springer Series. Margottini C, Canuti P, Sassa K, et al, editors. In: Proceedings of the Second World Landslide Forum; Rome (ITALY); 3–9 October 2011; 2013. pp. 671-678

[5] Segoni S, Piciullo L, Gariano SL. A review of the recent literature on rainfall thresholds for landslide occurrence. Landslides. 2018;**15**:1483-1501. DOI: 10.1007/s10346-018-0966-4

[6] Bogaard T, Greco R. Invited perspectives: Hydrological perspectives on precipitation intensity-duration thresholds for landslide initiation: Proposing hydro-meteorological thresholds. Natural Hazards and Earth System Sciences. 2018;**18**:31-39

[7] Crozier MJ. Prediction of rainfall-triggered landslides: A test of the antecedent water status model. Earth Surface Processes and Landforms. 1999; **24**:825-833

[8] Glade T, Crozier MJ, Smith P. Applying probability determination to refine landslide-triggering rainfall thresholds using an empirical "antecedent daily rainfall model". Pure and Applied Geophysics. 2000;**157**:1059-1079

[9] Godt JW, Baum RL, Chleborad AF. Rainfall characteristics for shallow landsliding in Seattle, Washington, USA. Earth Surface Processes and Landforms. 2006;**31**:97-110

[10] Brocca L, Ponziani F, Moramarco T, Melone F, Berni N, Wagner W. Improving landslide forecasting using ASCAT-derived soil moisture data: A case study of the Torgiovannetto landslide in Central Italy. Remote Sensing. 2012;**4**: 1232-1244. DOI: 10.3390/rs4051232

[11] Coe J. Regional moisture balance control of landslide motion: Implications for landslide forecasting in a changing climate. Geology. 2012;**40**(4):323-326. DOI: 10.1130/G32897.1

[12] Ponziani F, Pandolfo C, Stelluti M, Berni N, Brocca L, Moramarco T. Assessment of rainfall thresholds and soil moisture modeling for operational hydrogeological risk prevention in the Umbria region (Central Italy). Landslides. 2012;**9**:229-237. DOI: 10.1007/s10346-011-0287-3

[13] Mirus BB, Rachel I, Becker E, Rex I, Baum L, Joel I, et al. Integrating real-time subsurface hydrologic monitoring with empirical rainfall thresholds to improve landslide early warning. Landslides. 2018;**15**:1909-1919

[14] Mirus BB, Morphew MD, Smith JB. Developing hydro-meteorological tfor shallow landslide initiation and early warning. Water. 2018;**10**:1274. DOI: 10.3390/w10091274

[15] Valenzuela P, Domínguez-Cuesta MJ, Mora García MA, Jiménez-Sánchez M.

Rainfall thresholds for the triggering of landslides considering previous soil moisture conditions (Asturias, NW Spain). Landslides. 2018;**15**:273-282. DOI: 10.1007/s10346-017-0878-8

[16] Torres R, Dietrich WE, Montgomery DR, Anderson SP, Loague K. Unsaturated zone processes and the hydrological response of a steep unchanneled catchment. Water Resources Research. 1998;**34**(8): 1865-1879. DOI: 10.1029/98WR01140

[17] Baum RL, McKenna JP, Godt JW, Harp EL, McMullen SR. Hydrologic monitoring of landslide-prone coastal bluffs near Edmonds and Everett, Washington, 2001–2004: U.S. Geological Survey Open-File Report 2005–1063; 2005. p. 42

[18] Ebel BA, Loague K, Montgomery DR, Dietrich WE. Physics-based continuous simulation of long-term near-surface hydrologic response for the Coos Bay experimental catchment. Water Resources Research. 2008;**44**:W07417. DOI: 10.1029/2007WR006442

[19] Napolitano E, Fusco F, Baum RL, Godt JW, De Vita P. Effect of antecedent-hydrological conditions on rainfall triggering of debris flows in ash-fall pyroclastic mantled slopes of Campania (southern Italy). Landslides. 2015;**13**:967-983. DOI: 10.1007/s10346-015-0647-5

[20] Pieri P, Tropeano M, Sabato L, Lazzari M, Moretti M. Quadro stratigrafico dei depositi regressivi della Fossa bradanica (Pleistocene) nell'area compresa fra Venosa e il Mar Ionio. Giornale di Geologia. 1998:318-320

[21] Lazzari M, Pieri P. Modello stratigrafico-deposizionale della successione regressiva infrapleistocenica della Fossa bradanica nell'area compresa tra Lavello. Genzano e Spinazzola. Mem. Soc. Geol. It. 2002;**57**(1):231-237

[22] Piccarreta M, Pasini A, Capolongo D, Lazzari M. Changes in daily precipitation extremes in the Mediterranean from 1951 to 2010: The Basilicata region, southern Italy. International Journal of Climatology. 2013;**33**(15):3229-3248. DOI: 10.1002/joc.3670

[23] Lazzari M. Note illustrative della Carta Inventario delle Frane della Basilicata centroccidentale. Lagonegro: Editore Grafiche Zaccara; 2011. p. 136

[24] Lazzari M, Gioia D. Regional-scale landslide inventory, central-western sector of the Basilicata region (southern Apennines, Italy). Journal of Maps. Published Online. 2015;**12**(5):852-859. DOI: 10.1080/17445647.2015.1091749

[25] Lazzari M, Gioia D, Anzidei B. Landslide inventory of the Basilicata region (southern Italy). Journal of Maps. 2018;**14**(2):348-356. DOI: 10.1080/17445647.2018.1475309

[26] Brunetti MT, Peruccacci S, Rossi M, Luciani S, Valigi D, Guzzetti F. Rainfall thresholds for the possible occurrence of landslides in Italy. Natural Hazards and Earth System Sciences. 2010;**10**:447-458

[27] Vennari C, Gariano SL, Antronico L, Brunetti MT, Iovine G, Peruccacci S, et al. Rainfall thresholds for shallow landslide occurrence in Calabria, southern Italy. Natural Hazards and Earth System Sciences. 2014;**14**:317-330. DOI: 10.5194/nhess-14-317-2014

[28] Piciullo L, Gariano SL, Melillo M, Brunetti MT, Peruccacci S, Guzzetti F, et al. Definition and performance of a threshold-based regional early warning model for rainfall-induced landslides. Landslides. 2016b;**14**:995-1008. DOI: 10.1007/s10346-016-0750-2

[29] Peruccacci S, Brunetti MT, Gariano SL, Melillo M, Rossi M, Guzzetti F, et al. Rainfall thresholds for possible landslide occurrence in Italy. Geomorphology. 2017;**290**:39-57. DOI: 10.1016/j.geomorph.2017.03.031

[30] Ray RL, Jacobs JM. Relationships among remotely sensed soil moisture, precipitation and landslide events. Natural Hazards. 2007;**43**:211-222

[31] Basilicata Region, Ufficio Produzioni Vegetali e Silvicoltura Produttiva - Dipartimento Agricoltura, Sviluppo Rurale, Economia Montana, Carta Pedologica della Basilicata—S.EL. CA srl—Firenze. 2006. Available from: http://www.basilicatanet.it/suoli/cartine/carta-pedologica.zip http://dati.regione.basilicata.it/catalog/dataset/carta-pedologica ttp://www.soilmaps.it/download/csi-BrochureSR_a4.pdf

[32] Corine Land Cover. 2012. Available from: http://www.sinanet.isprambiente.it/it/sia-ispra/download-mais/corine-land-cover/corine-land-cover-2012/view

[33] Ruiz-Pérez G, Koch J, Manfreda S, Caylor KK, Francés F. Calibration of a parsimonious distributed ecohydrological daily model in a data scarce basin using exclusively the spatio-temporal variation of NDVI. Hydrology and Earth System Sciences. 2017;**21**:6235-6251. DOI: 10.5194/hess-21-6235-2017

[34] Manfreda S, Mita L, Dal Sasso SF, Samela C, Mancusi L. Exploiting the use of physical information for the calibration of the lumped hydrological model. Hydrological Processes. 2018;**32**(10):1420-1433. DOI: 10.1002/hyp.11501

[35] Farmer DL, Sivapalan M, Jothityangkoon C. Climate, soil and vegetation controls upon the variability of water balance in temperate and semi-arid landscapes: Downward approach to hydrological prediction. Water Resources Research. 2003;**39**:2

[36] De Smedt F, Yongbo L, Gebremeskel S. Hydrologic modelling on a catchment scale using GIS and remote sensed land use information. WIT Transactions on Ecology and the Environment. 2000;**45**:295-304

[37] Liu YB, Gebremeskel S, De Smedt F, Hoffmann L, Pfister L. A diffusive transport approach for flow routing in GIS-based flood modeling. Journal of Hydrology. 2003;**283**(1–4):91-106

[38] Manfreda S, Fiorentino M, Iacobellis V. DREAM: A distributed model for runoff, evapotranspiration, and antecedent soil moisture simulation. Advances in Geosciences. 2005;**2**:31-39. DOI: 10.5194/adgeo-2-31-2005

[39] Eagleson PS et al. Climate, soil, and vegetation. Climate, soil, and vegetation. Water Resources Research. 1978;**14**(5):705-776

[40] Rosano R, Manfreda S, Fiorentino M, Sole A. Sviluppo ed Ingegnerizzazione di un Applicativo Software per la Modellazione Idrologica a Scala di Bacino, 29° Convegno di Idraulica e Costruzioni Idrauliche: Editoriale Bios; 2008

[41] Manfreda S, Brocca L, Moramarco T, Melone F, Sheffield J. A physically based approach for the estimation of root-zone soil moisture from surface measurements. Hydrology and Earth System Sciences. 2014;**18**:1199-1212. DOI: 10.5194/hess-18-1199-2014

[42] Lazzari M, Piccarreta M, Manfreda S. The role of antecedent soil moisture conditions on rainfall-triggered shallow landslides. Natural Hazards Earth System Science Discussion. 2019:1-11. DOI: 10.5194/nhess-2018-371

[43] Baum RL, Godt JW. Early warning of rainfall-induced shallow landslides and debris flows in the USA. Landslides. 2009;7(3):259-272

[44] Baldwin D, Manfreda S, Keller K, Smithwick EAH. Predicting root zone soil moisture with soil properties and satellite near-surface moisture data at locations across the United States. Journal of Hydrology. 2017;**546**:393-404. DOI: 10.1016/j.jhydrol.2017.01.020

Chapter 6

Long-Term Monitoring of Slope Movements with Time-Domain Reflectometry Technology in Landslide Areas, Taiwan

Miau-Bin Su, I-Hui Chen, Shei-Chen Ho, Yu-Shu Lin and Jun-Yang Chen

Abstract

The study employs time-domain reflectometry (TDR) technology for landslide monitoring to explore rock deformation mechanism and to estimate locations of potential sliding surfaces in several landslide areas, Taiwan, over ten years. Comparing to laboratory and field testing, sliding surfaces in landslide areas occurred mainly at two types, namely shear and extension failure. The TDR technology is used for field monitoring to analyze locations of sliding surfaces and to quantify the magnitude of the sliding through laboratory shear and extension tests. There are several TDR-monitoring stations in six alpine landslide areas in the middle of Taiwan for long-term monitoring. A relation between TDR reflection coefficients and shear displacements was employed for a localized shear deformation in the field. Furthermore, the type of a cable rupture for the TDR monitoring in landslides can be determined as shear, extension, or compound failure through the field TDR waveforms. Overall, the TDR technology is practically used for a long-term monitoring system to detect the location and magnitude of slope movement in landslide areas.

Keywords: time domain reflectometry (TDR), landslide monitoring, slope movement, shear and extension testing, deformation quantification

1. Introduction

Determining the location and magnitude of sliding surfaces is a vital measure for landslide monitoring. For conventional slope monitoring, drill-log reports can illustrate in situ presence of weak rock or location of weak rock masses [1, 2]. Furthermore, inclinometers are used for landslide monitoring to detect sliding zones and to measure the subsurface lateral displacement of soil or rock [3–5]. However, it is time-consuming and difficult by these traditional methods to interpret an accurate location of sliding surface in a landslide area [6, 7]. In recent years, time-domain reflectometry (TDR) is employed for the monitoring of slope movement to locate depths of slope failures [8, 9]. The TDR technology uses a cable tester to detect a coaxial cable grouted in a borehole. While the cable broken or ruptured, a TDR signal from the cable tester is reflected. The reflection shows the location of sliding surfaces and the effect of TDR-cable length can be determined accurately [10–13].

There are some characteristics of the TDR technology for the monitoring of landslide deformation. Firstly, a coaxial cable is a continuous sensor where the TDR system can interrogate reflection coefficients of a crimp, kink, extension, or break [14]. Secondly, an important ability for the TDR technology is to locate the relative movement along the cable by extended or shear stress that causes the change of cable characteristic impedance when two adjacent rock masses move each other [15, 16]. Finally, reflection coefficients of TDR-cable deformation are positive correlations with the rate of slope movement [9, 17, 18].

There are two main themes in this research. The first is for laboratory experiments that include shear and extension tests with grouted cables. The TDR technology can interrogate reflected waveforms corresponding to the cable deformation in order to get a relation between reflection coefficients and the magnitude of cable deformation [17]. The other theme is for field TRD monitoring. There are several TDR monitoring stations in six landslide areas where occurred some slides in Taiwan [6, 17–19]. Finally, the study analyzes field data from TDR monitoring stations to compare to results of quantification in laboratory tests and to determine the location, type, and magnitude of slope movement for long-term monitoring in the landslide areas.

2. Laboratory testing for TDR technology

For literature reviews of TDR laboratory experiments, a TDR tester detected reflection coefficients of a grouted coaxial cable when the cable is deformed by shear or extension stress as shown in **Figure 1(a)**, which illustrated a signal “spike” as like a shape of “V” by shear stress (**Figure 1(b)**) and another signal as like a shape of “concave” by extension stress (**Figure 1(c)**) based on the TDR theories [15, 16].

Whether shearing deformations are big or small, the magnitude of TDR reflected waveforms is significant correlation with cable deformation [20, 21]. An integrated area method is proposed to quantify the magnitude of a TDR reflected waveform on the deformation of rock masses and concrete structures through laboratory bench tests [10, 17]. The method involved the integration of the TDR voltage signal over time in **Figure 2**, which means that L1 is the start point of a TDR reflected waveform due to a change in capacitance and L2 is an end point when the voltage returns to its reference level [17].

For laboratory tests, each coaxial cable was grouted into a cement test specimen, which was 1-m in length and 10 cm in diameter. Then, each test specimen was installed on a bench, as shown in **Figures 3** and **4**, which were twofold. One is to determine a relation between the magnitude of cable deformations and TDR reflected waveforms by shear displacement. The second is to measure a change length of the cable with TDR reflected waveforms by extension displacement.

The laboratory testing was designed to define a relation between the TDR reflection coefficients and the magnitude of cable deformation by shear and extension failure. For example, a coaxial cable was grouted into a cement test specimen with 1 m in length and then connected to a TDR cable tester (HL 1101). The test specimen was installed on a bench, which was divided into three 20-cm-length and 10-cm-diameter units, named S1, S2, and S3, as shown in **Figure 3**. Then, a TDR tester (HL1101) sent a voltage pulse waveform that travels along a coaxial cable. Finally, data from the tester were transmitted to a PC and was displayed as TDR waveforms. The cable and grout were the same as the materials used in field installation, listed in **Table 1** [17].

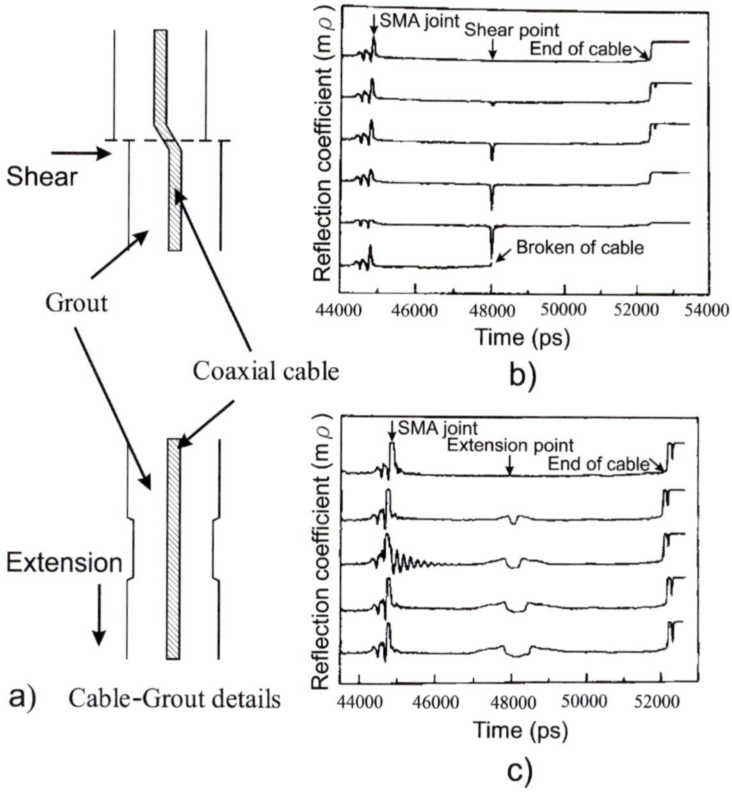

Figure 1.
Diagrams of reflected waveforms of TDR coaxial cables in grout by shear and extension movement. (a) grouted cables deformed by shear or extension stress; (b) changes in reflection coefficients of TDR waveforms for shear tests; (c) changes in reflection coefficients of TDR waveforms for extension tests [15, 16].

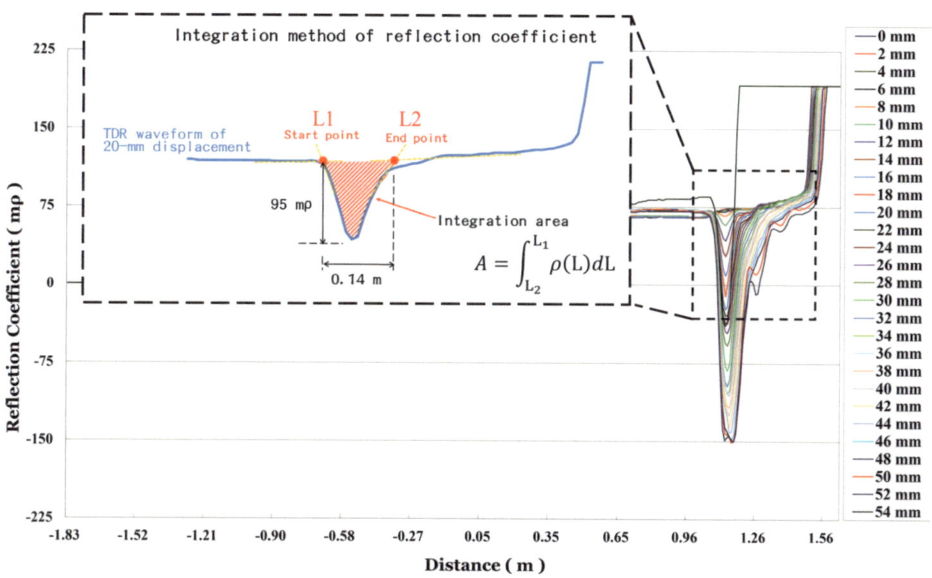

Figure 2.
Integration method of the TDR waveform for shear displacement [17].

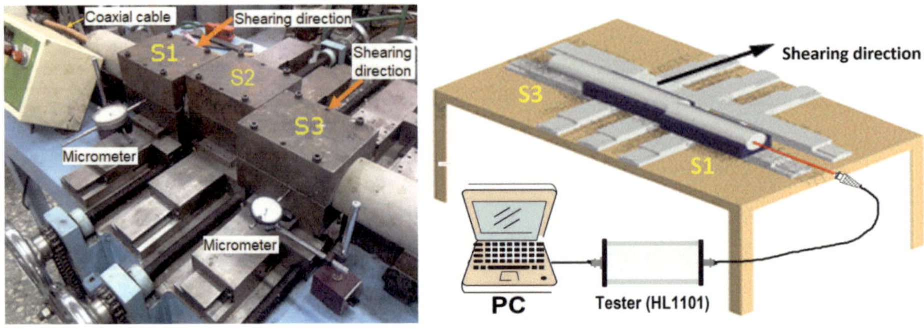

Figure 3.
Instrumentation of shear bench test (adapted from [17]).

Figure 4.
Instrumentation of extension bench test (adapted from [18]).

Coaxial cable Type	Diameter (mm)	Grout cement Type	Grout mix ratio (water to cement)	Compressive strength (kgf/cm^2)
CTLLCX 7/8″ CFC	Φ25 mm	Portland cement I	1:3	146.5

Table 1.
Properties of the TDR cable and grout [17].

For the shear test, S1 and S3 were shear units, but S2 was a fixed unit in **Figure 3**. The S1 was set at a given shear displacement of 0, 10, 20, 30, and 40 mm, and then the S3 was sheared successively at 2-mm intervals for each shear displacement of the S1, respectively, until the cable rupture [17]. Results of the TDR interrogated waveforms for the cable were plotted in **Figure 5(a)** where the TDR signals (reflection coefficient in millirhos) represented the amplitude of reflected waveforms at a distance in meter along the cable [17]. As can be seen in **Figure 5(b)**, the shear displacement at the S3 location is 52 mm before the cable rupture corresponding to the reflection coefficient at 233 mρ.

All shear tests have the same tendencies as the graph of **Figure 5** for the TDR responses of the S1 and S3 displacement, so there existed a significant relation between shear displacements and reflection coefficients, as shown in **Figure 6**, which illustrated the maximum and average magnitude of the cable deformation by shear failure that were 60 and 47 mm, respectively [17]. A linear relationship between shear deformation and TDR reflection magnitude is attractive due to its simplicity, which has been used extensively to quantify the magnitude of shear deformation [8]. Finally, the study uses the linear regression to determine the

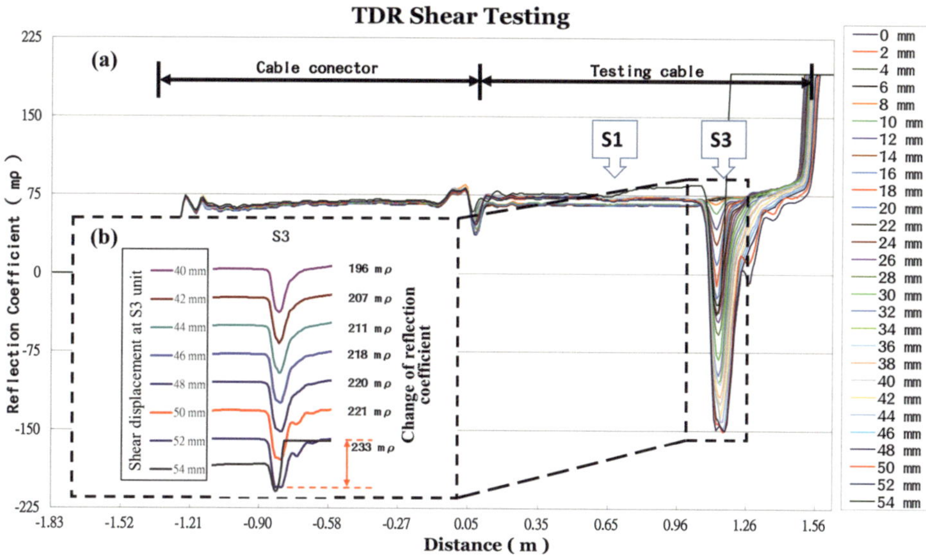

Figure 5.
Graph of reflected waveform changes at S3 shear plane as S1 shear displacement is set at 0 mm. (a) changes in reflection coefficients of TDR waveforms for the whole grouted cable; (b) enlarged graph of changes in reflection coefficients of TDR waveforms at S3 shear plane [17].

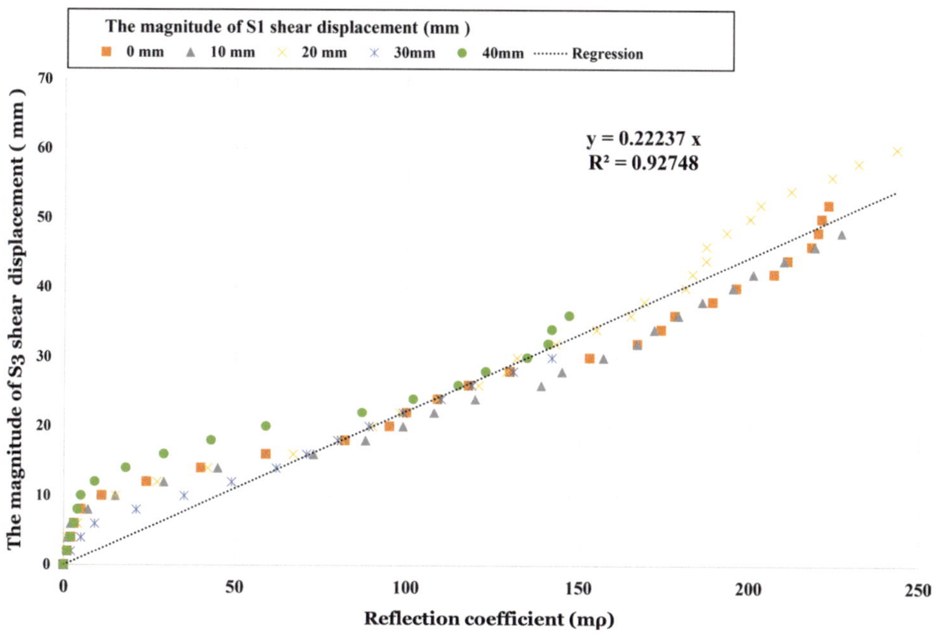

Figure 6.
Graph of the relationship between the shear displacement (millimeters) and reflection coefficient (mρ) in shear tests [17, 18].

magnitude of shear deformations in the field corresponding to relative changes of reflection coefficients interrogated by on-site TDR system [17, 18].

For the extension test, a 1-m test specimen was installed on the bench, which was set as two 20-cm-length units; one was a moving unit T1 and the other was fixed, as shown in **Figure 4** [18]. The amount of extension displacement at the T1 location successively incremented at 5 mm until the cable ruptured and results

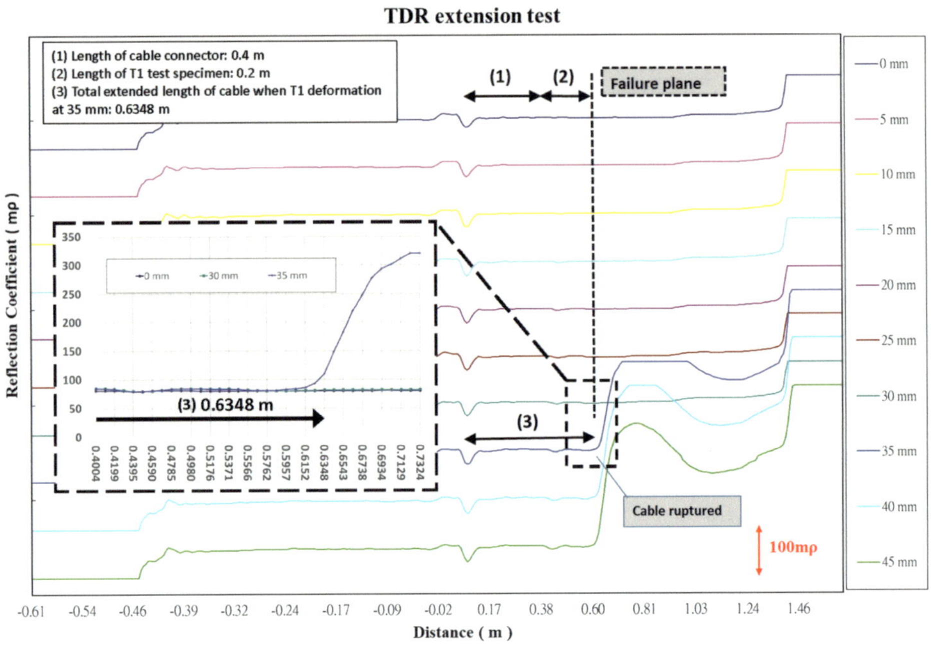

Figure 7.
Graph of reflected waveform changes in extension tests [18].

of the extension test was plotted in **Figure 7**, which shows that the magnitude of extension displacement is 35 mm as the cable rupture [18]. A change in length of the cable between first waveform and the 35-mm-test waveform was measured to 0.0348 m that was almost the same as the magnitude of extension displacement at 35 mm. Thus, the result can be employed to measure the magnitude of extended cable in field TDR monitoring [6, 18].

3. Principle of the TDR monitoring in landslide areas

A diagram of on-site TDR instrumentation is shown in **Figure 8**. A coaxial cable is grouted with cement into a borehole in a slope. A TDR cable tester (HL1101) sends and receives a voltage pulse waveform that travels along the coaxial cable through a connector. A transmission of the TDR signal through the tester is stored in a microcomputer that is an embedded system after the signal is reflected by a cable deformation.

There are six TDR-monitoring stations in different landslide areas, Taiwan, including Lishan, Song-Mao, Old Tribe, and New Jia-Yang areas since 2008, and Jiufenershan and Lushan areas since 2012 and 2018, respectively. For the first four areas, the distribution of TDR-monitoring stations was shown in **Figure 9**, which illustrated representative stations called B5, L2, J1, and S9 monitoring stations in the four areas. The four landslide areas are located at the middle of Taiwan with altitude between 1800 and 2100 m, inclination toward the northwest and slope angle of 15–30° [22]. Geologically, the areas belong to colluvium and slate formation. In April 1990, a severe landslide occurred in Lishan area after prolonged torrential rainfall. After that, governments remedied the area with many drainage works and slope monitoring systems in order to prevent from disasters again [6].

The TDR cable of 40–50 m in depth was grouted into a borehole every station. All TDR waveforms of different date are recorded every month so that it is easy

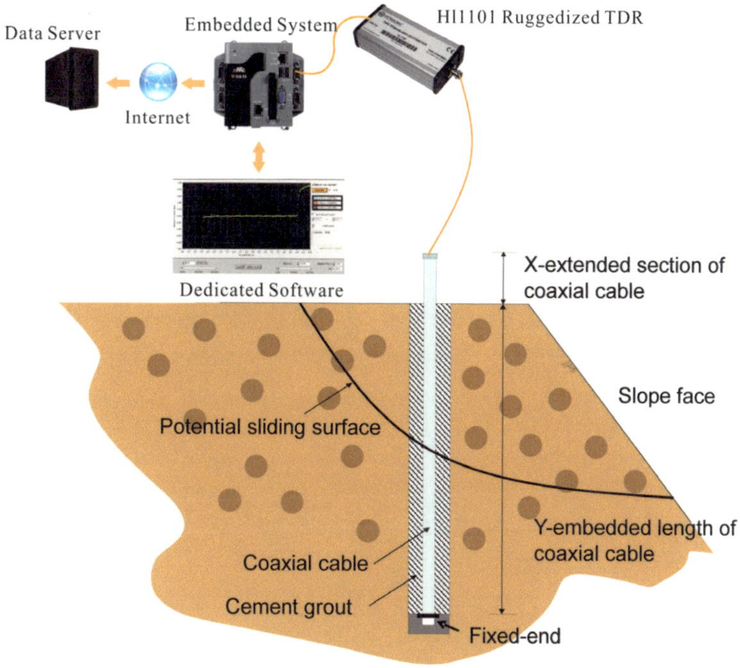

Figure 8.
Diagram of on-site instrumentation for the TDR technology (adapted from [18]).

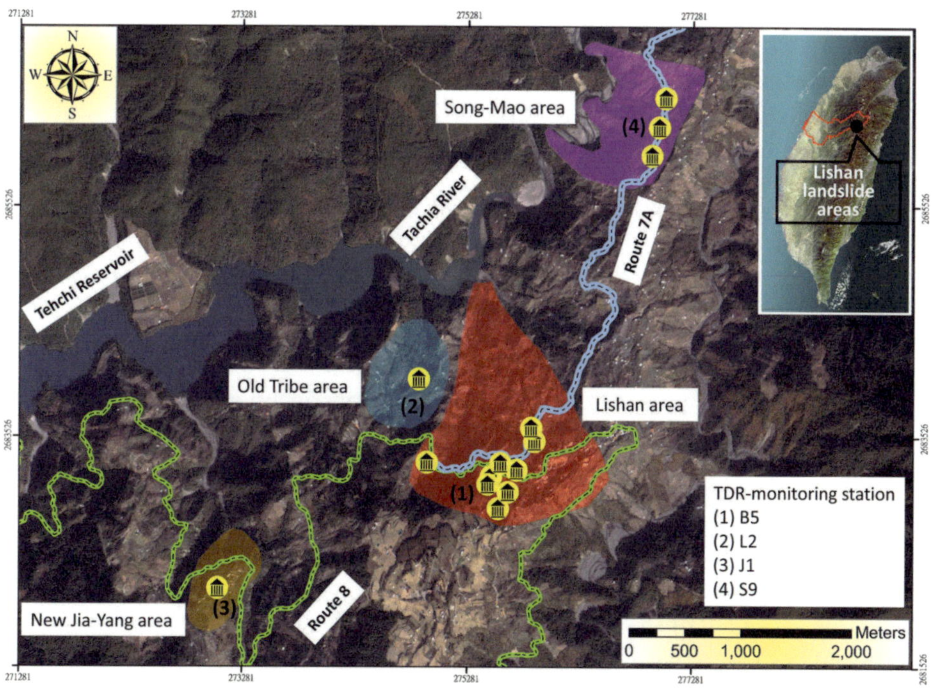

Figure 9.
Location map of TDR monitoring stations in four landslide areas.

to interpret a change of waveforms at any depth where a slope movement could occur. For example, records of the TDR monitoring at J1 TDR-monitoring station in New Jia-Yang landslide area were plotted in **Figure 10**. They revealed changes in

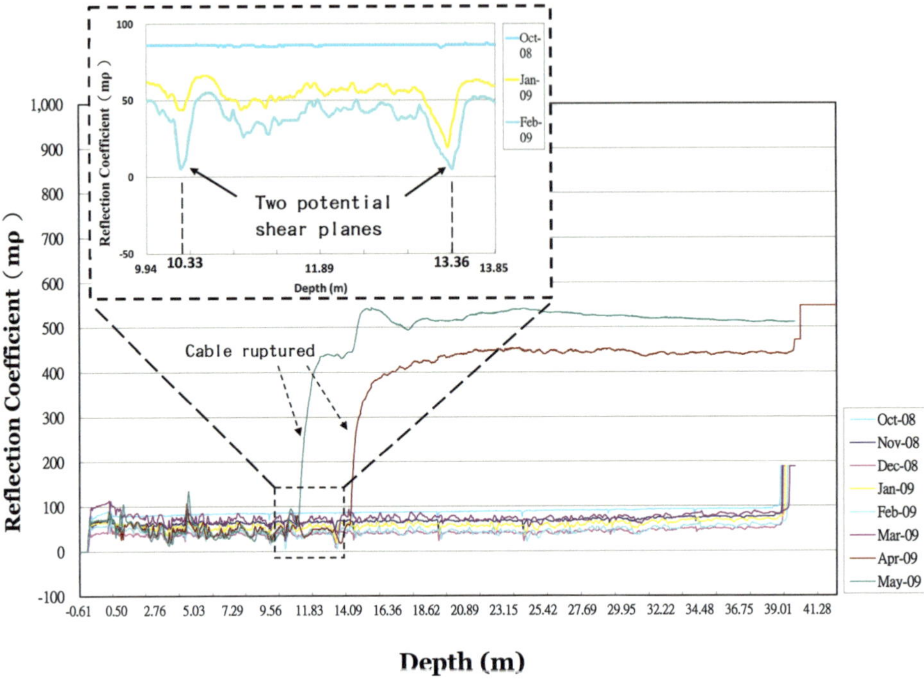

Figure 10.
Recorded waveform at J1 TDR-monitoring station [17].

reflection coefficients at the J1 station since 2008. The changes in TDR waveforms from 10/2008 to 05/2009 shown in the enlarged graph of **Figure 10** were caused by two spikes at 10.33 and 13.36 m in depth at the station, respectively [17]. Thus, significant increases of amplitudes in the reflection coefficient occurred at the two depths in April and May 2009, indicating that the cable was ruptured at two locations by shear failure [6, 17].

The other case study is in Jiufenershan landslide area that is located at the middle of in Taiwan, as shown in **Figure 11**. The area occurred due to a severe landslide triggered as a result of the 1999 earthquake. The geology of the landslide area with altitude between 500 and 2100 m is underlain mainly by Miocene sedimentary and shale formations where the strike is N36°E and the dip is 21°SE [23]. Since 2003, there has been a landslide monitoring project for the Jiufenershan landslide with some monitoring equipment, including extensometers, inclinometers, groundwater level gauges, etc. The TDR technology was installed in the area since 2012 in order to monitor slope movement in the area [18].

From the 12/2012 and 01/2013 TDR records in Jiufenershan landslide area (**Figure 12**), there were significant deformations at 7.51- and 35.72-m depths at N2 station because there exited localized cable ruptures [18]. Based on the regression of the relation between cable deformation and reflection coefficient in the laboratory, 47-mm displacement of the cable corresponds to approximately the TDR reflection coefficient of 210 mρ that can be regarded as the working limit for the cable rupture by deformed failure in the case study [18].

For the last case study in Lushan landslide area, it is located in central Taiwan with an altitude from 1050 to 1480 m, as shown in **Figure 13** [19]. The bedrock of the slope mass is mainly composed of slate with strikes of cleavage range from N10° to 40°E. The slope is defined as composed of 20-m-thick colluvium near the

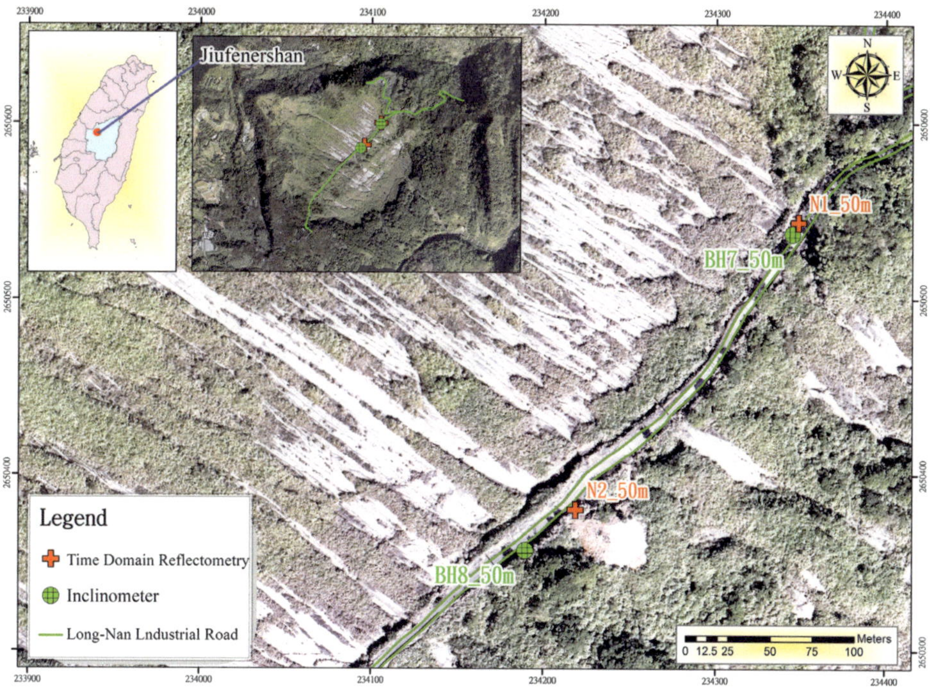

Figure 11.
Location map of the monitoring stations in Jiufenershan landslide area [18].

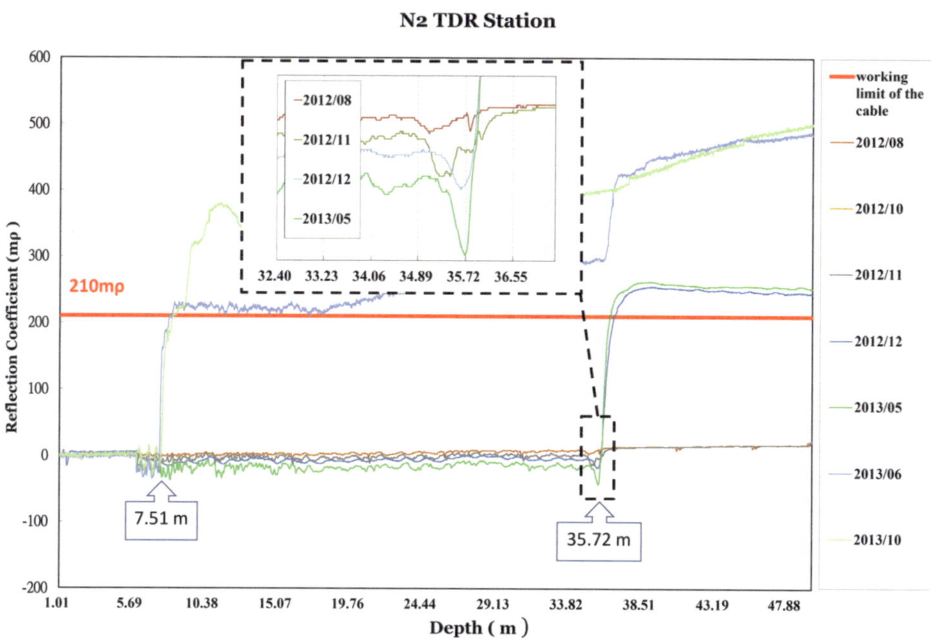

Figure 12.
Changes of reflection coefficients interrogated by the TDR at N2 monitoring station [18].

surface and fresh slate at bottom [24]. The Lushan formation in the area belongs to the attitude of the foliation, which influences the mass rock creep structures of foliated rocks and may experience deformation by potential landslides [25]. There

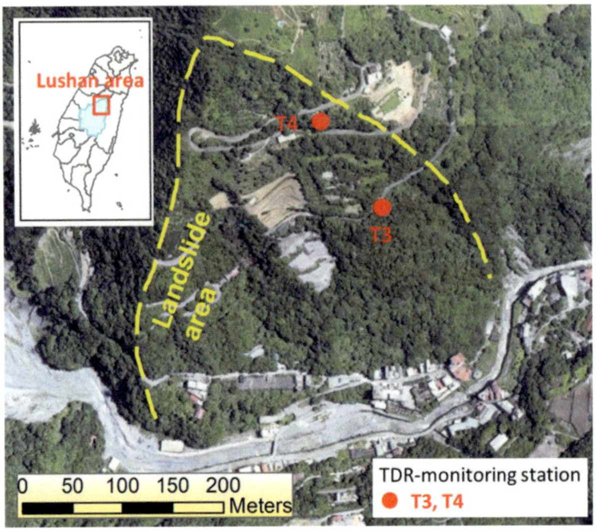

Figure 13.
Location and geologic map in Lushan landslide area.

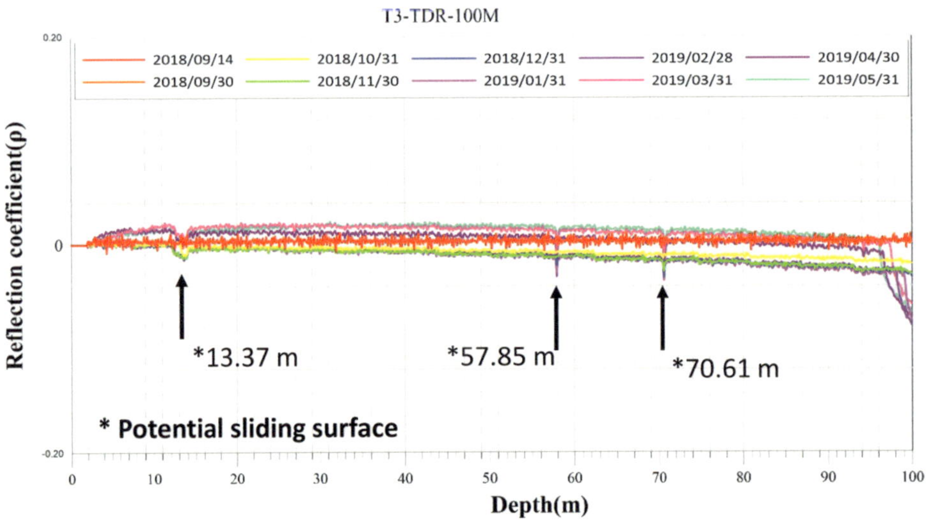

Figure 14.
Records of TDR reflected waveforms at T3 monitoring station (adapted from [19]).

are two TDR-monitoring stations, coded T3 and T4, in the area where coaxial cables of approximately 100-m length were grouted with cement to monitor potential slope movement.

For example, records of T3 TDR-monitoring waveforms from Sep 2018 to May 2019 were plotted in **Figure 14**. Changes in TDR waveforms were caused by three spikes at approximately 13, 57, and 70 m in depth at the T3 station [19]. These TDR signals "spikes" can be indicated as the location of localized shear failure through the laboratory testing in the study. Furthermore, significant increases in amplitudes in the reflection coefficient occurred at the three depths from September 2018 to May 2019, indicating that the cable was partially broken at the three locations by shear deformation at T3 [19].

4. Results and discussions of field monitoring

According to the result of laboratory tests and the data from field monitoring, the type and magnitude of slope movement in the landslide areas can be determined. All recorded waveforms of different dates were separately plotted in a graph to easily compare changes in waveforms of the TDR monitoring. Thus, records of the TDR waveforms in the landslide areas were replotted in **Figures 15–18** that showed changes in reflection coefficients of the four stations (B5 in Lishan area, J1 in New Jia-Yang area, L2 in Old Tribe area, and S9 in Song-Mao area). TDR-cable signals presented the amplitude of reflected waveforms at a given distance corresponding to the depth of a grouted cable at TDR-monitoring stations. Then, the waveforms can be easily interpreted by a change of waveforms at a depth where shear failure occurred or at an extended length of the grouted cable where extension failure occurred.

As illustrated in **Figure 15**, no significant change in the waveforms occurred at the B5 station from 2016 to 2018 because remedial works had been performed to prevent the area from additional sliding. Alternatively, some TDR waveforms changed at the other three stations from 2008 to 2015 in **Figures 16–18**, which illustrated that the type of waveform changes was shear displacement and extension signals.

For J1 TDR-monitoring station, the TDR waveforms were formed as like the shape of "V" at a depth of approximately 13 m in **Figure 16**, which illustrated the type of the waveform by shear stress [17]. Based on the laboratory tests, the magnitude of the shear displacement in **Figure 16** was more than 47 mm in July 2009 because the reflection coefficient in the spike of the TDR waveform was over 210 mρ that means the localized cable ruptured.

From another point of view in other graphs, some TDR waveforms of other areas were formed as the type of extension failure such as L2 and S9 stations because the type of the waveforms was similar to the type of the laboratory extension testing and total length of the two cables was extended at the end of them. As shown by the Mar/2009 TDR record in **Figure 17**, there was an extended displacement at a

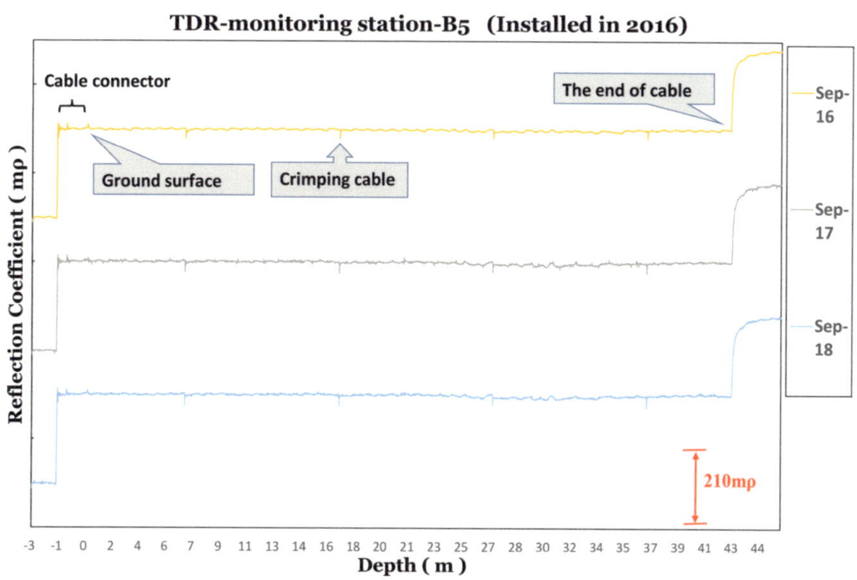

Figure 15.
Graph of TDR reflected waveforms at B5 TDR-monitoring station.

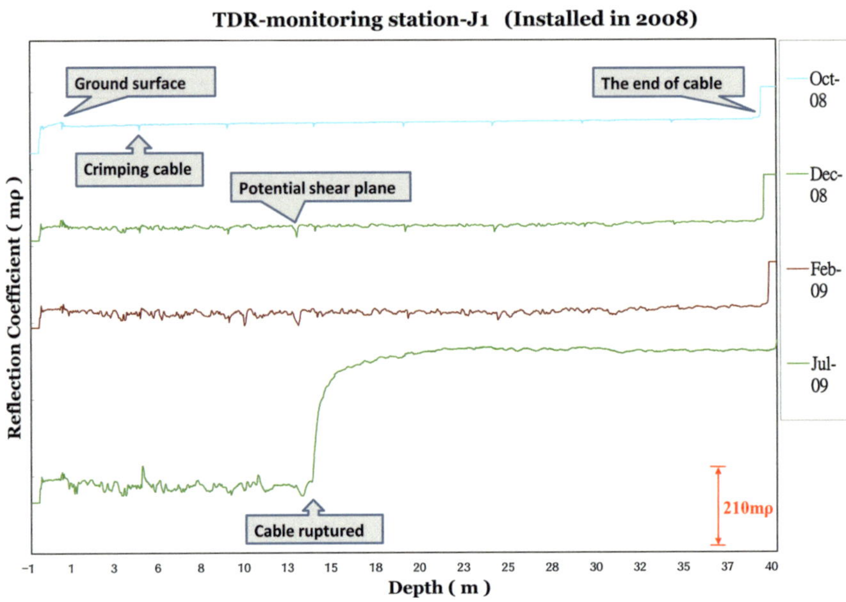

Figure 16.
Graph of TDR reflected at J1 TDR-monitoring station.

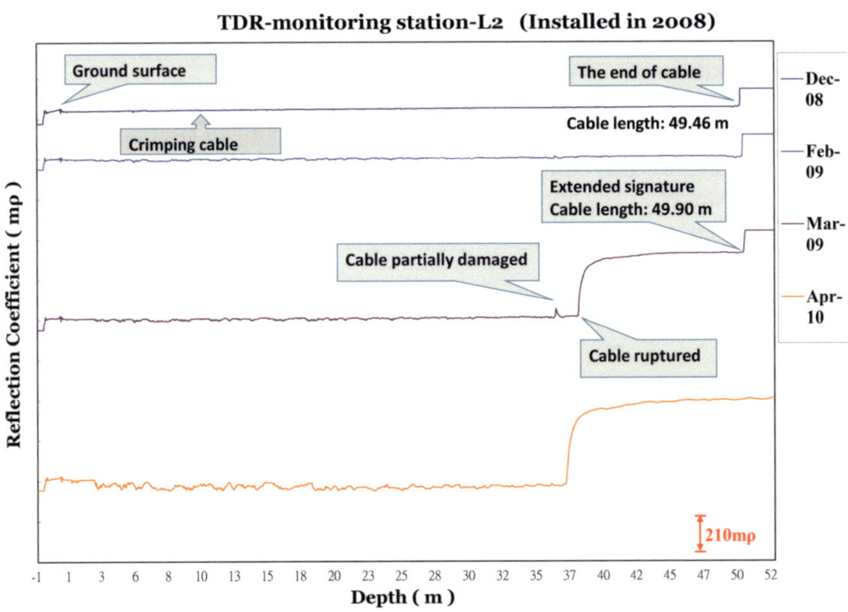

Figure 17.
Graph of TDR reflected at L2 TDR-monitoring station.

depth of 37.87 m at L2 station when the waveform signal showed the cable ruptured. As can be seen from **Figure 18**, the cable was extended to a depth of 43.23 m at S9 station in October 2014. Also, there were considerably extended lengths of the end cables at L2 and S9 stations, by 0.44 and 0.20 m, respectively.

For N2 TDR-monitoring station in Jiufenershan landslide area, records of on-site TDR monitoring waveforms were replotted in **Figure 19** that showed the TDR waveforms were formed as signals of cable rupture at 7.51- and 35.72-m depth marked as A and B, respectively [18]. As shown by the 11/2012 TDR record of the enlarged

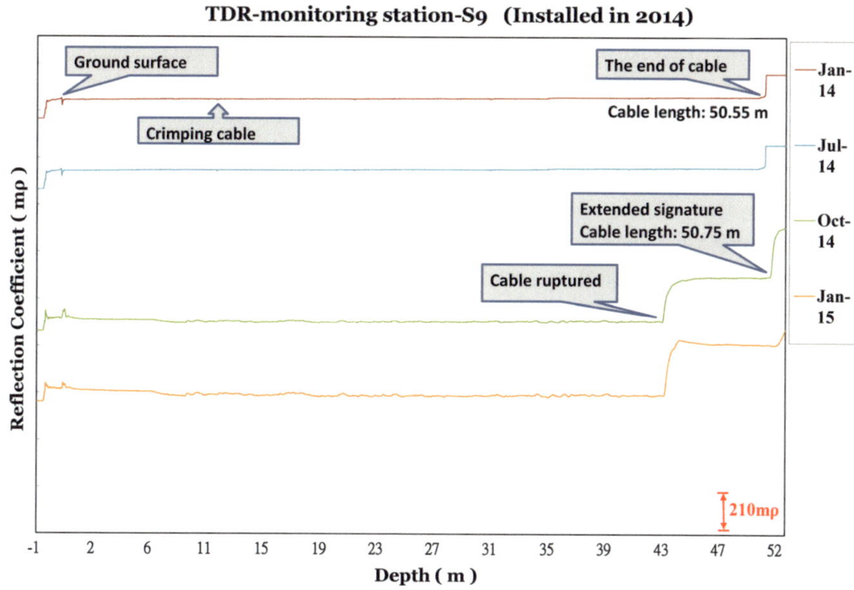

Figure 18.
Graph of TDR reflected at S9 TDR-monitoring station.

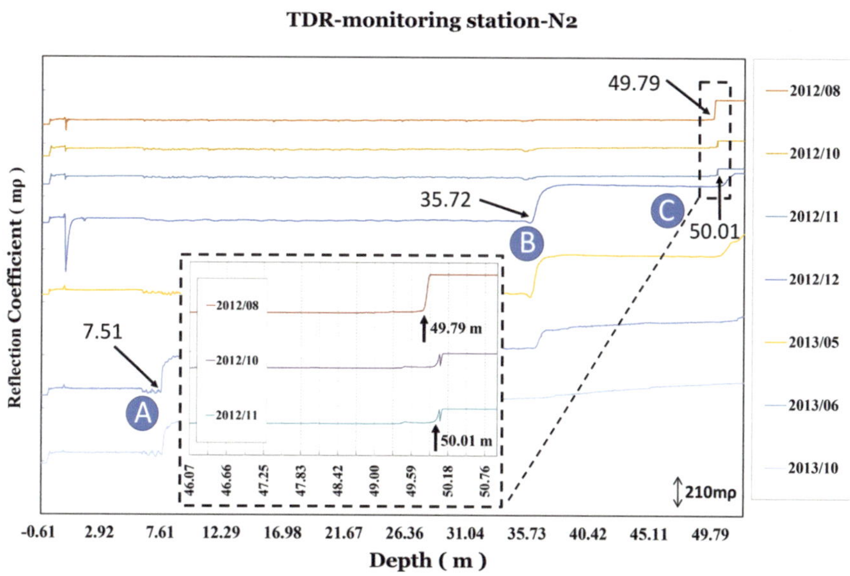

Figure 19.
Graph of TDR reflected at N2 TDR-monitoring station [18].

chart in **Figure 19** (mark C), the total length of the cable had been extended by potential extension stress from 49.79 to 50.01 m measured by the TDR signal at the end of the cable. It was obvious that the cable was deformed by extension stress initially [18]. However, for the enlarged graph in **Figure 12**, there was a pulse-like signal of the waveform at a 35.72-m depth in November 2012 that was similar to a spike type generated by laboratory shear testing; thus, the cable was deformed by shear stress at the location before the cable rupture in December 2012, despite the fact that the cable had been extended from August 2012 to November 2012. As a result, the cable would be damaged by compound failure in the landslide area [18].

For T3 TDR-monitoring station in Lushan landslide area, original records of TDR-monitoring waveforms were replotted as a graph rotated vertically so that it is easy to observe changes of TDR waveforms in vertical depth under the ground at the T3 station, as shown in **Figure 20** where x axis means the T3 shear displacement in centimeters calculated using the linear regression equation in the laboratory testing [18, 19].

There were three shear surfaces at approximately 13-, 57- and 70-m depths in **Figure 20** as well as in **Figure 14**. Through the laboratory testing, the magnitude of cable sheared deformation can be determined for the field slope movement corresponding to the TDR reflection coefficients in the T3 TDR monitoring. For example, increases in the amplitudes of reflection coefficients occurred at two spikes at approximately 57- and 70 m depths from September 2018 to May 2019 in the enlarged graph of **Figure 20**. The magnitude of the cable sheared deformation at the two depths was calculated using the linear regression equation $y = 0.22247x$ (see **Figure 6**, where y means shear displacement in millimeters and x means a change of reflection coefficients in mρ; [18, 19]). Then, the changes in reflection coefficients of the two localized shear locations were detected as 0.03 and 0.024 ρ (30 and 24 mρ) corresponding to shear displacement by 6.7 and 5.6 mm, respectively, on May 31, 2019. It was obvious that these TDR signal "spikes" can be indicated as the type of localized shear failure through the laboratory testing in the study. Meanwhile,

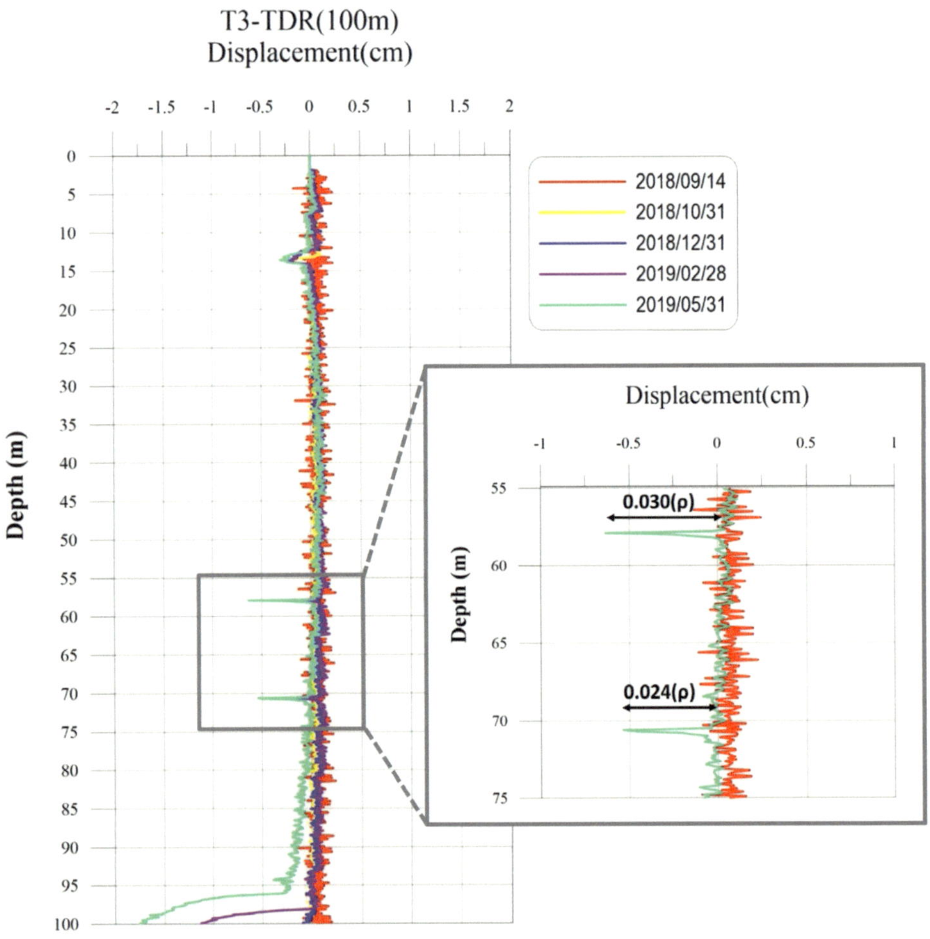

Figure 20.
Graph of TDR reflected waveforms at T3 TDR-monitoring station.

Area	Monitoring station	Time	Failure depth (m)	Failure type	Magnitude (m)
Lishan	B5	September 2015 to September 2017	No significant deformation		
New Jia-Yang	J1	May 2009	10.33	Shear	More than 0.047
		April 2009	13.36	Shear	More than 0.047
Old Tribe	L2	March 2009	37.87	Extension	Extended length: 0.44
Song-Mao	S9	October 2014	43.23	Extension	Extended length: 0.20
Jiufenershan	N2	December 2012	35.72	Coexistence of extension and shear type	Extended length: 0.22
		June 2013	7.51		
Lushan	T3	May 2019	13.37	Shear	0.0069
		May 2019	57.85	Shear	0.0067
		May 2019	70.61	Shear	0.0056

Table 2.
Location and magnitude of shear deformation in landslide areas, Taiwan.

reflection coefficients of the T3 TDR monitoring did not reach 210 mρ so the cable did not rupture. However, significant increases of amplitudes in the reflection coefficient occurred at three depths from September 2018 to May 2019 so that the area could consecutively occur as mass rock creep in Lushan landslide area [19].

By the method of the linear regression calculation, all magnitude of localized shear displacements at sliding surfaces at every TDR-monitoring station in the six landslide areas can be determined, as presented in **Table 2**. On the other hand, all magnitude of extended cables at L2, S9, and N2 stations can also be measured as shown in **Table 2**. In particular, the data from the N2 station inferred that sliding deformation in the landslide area would be a compound type that means the slope movement may be a coexistence of extension-type and shear-type deformation through the result of field TDR waveforms [18]. The magnitude of extension deformation was 0.22 m from August 2012 to November 2012 at N2 station. Also, the cable of N2 had been ruptured because the reflection coefficient of the TDR waveforms had been over 210 mρ (see **Figure 12**) according to the result of the laboratory testing.

Overall, if a reflection coefficient of a TDR monitoring station reaches to 210 mρ or more in the field, it could mean that a TDR cable could be damaged by potential slope deformation through the result of the laboratory and field tests. Average 47 mm of shear displacement in a TDR cable corresponds to the TDR reflection coefficient of 210 mρ that can be defined as the working limit for the cable ruptured by shear failure in these case studies. Nevertheless, the maximum amount of cable extension can reach 22 and 44 cm in Jiufenershan and Old Tribe landslide area, respectively. Finally, the TDR technology can determine the type, location, and magnitude of slope movement in the landslide areas although field TDR waveforms were complicated and diverse. The TDR technology is practical and successful for landslide monitoring to interpret the location and magnitude of sliding surfaces in the case studies.

5. Conclusions

The study employed the TDR technology for the landslide monitoring of slope movement to determine the location and magnitude of sliding surfaces in six landslide areas. Comparing to laboratory and field tests, the TDR monitoring of J1 in New Jia-Yang landslide area detected two sliding surfaces at 10.33- and 13.36-m depths where the magnitude of sliding deformation was estimated at more than 47 mm, because the reflection coefficients of the TDR monitoring were over 210 mρ. For the TDR monitoring of L2 and S9, the extended length of each grouted cable can be measured by 44 and 20 cm, while the two cables had been ruptured at 37.87- and 43.23-m depth, respectively. For the TDR monitoring of J1, there were cable ruptures at depths of 7.51 and 35.72 m where rock mass slides was extension-type and shear-type compound failure. Finally, the TDR monitoring of T3 consecutively detected three sliding surfaces at 13.37-, 57.85- and 70.61-m depths where slight deformation amount was estimated by 0.31, 0.67, and 0.53 cm, respectively, using the linear regression equation of the relation between TDR reflection coefficients and shear displacements in the shear tests.

Overall, the TDR technology can accurately detect the location of TDR cable deformation for the monitoring of slope movement in the landslide areas. Meanwhile, sliding types of shear, extension, or compound failure can be determined using the change of TDR reflected waveforms in the laboratory and field. Finally, 210-mρ reflection coefficients of TDR monitoring can be a threshold of a partial cable ruptured corresponding to the cable sheared deformation of 47 mm in the case studies. In the future, the TDR technology can be an early warning system before a landslide occurs.

Acknowledgements

The field test presented in this paper could be realized through the support and sponsorship of the Soil and Water Conservation Bureau (SWCB), Council of Agriculture, Executive Yuan, Taiwan. Consecutive projects including long-term monitoring are proposed and approved by the technical counseling committee on the remediation works of large-scale landslide areas, the SWCB. Furthermore, the performance evaluation of each project is periodically reviewed by the committee annually.

Author details

Miau-Bin Su, I-Hui Chen*, Shei-Chen Ho, Yu-Shu Lin and Jun-Yang Chen
Department of Civil Engineering, National Chung Hsing University, Taichung City, Taiwan

*Address all correspondence to: ian.cih82@gmail.com

References

[1] Goodman RE. Introduction to Rock Mechanics. 2nd ed. New York: Wiley; 1989

[2] Blackburn JT, Dowding CH. Finite-element analysis of time domain reflectometry cable-grout-soil interaction. Journal of Geotechnical and Geoenvironmental Engineering. 2004;**130**(3):231-239

[3] Dowding CH, Su MB, O'Connor KM. Measurement of rock mass deformation with grouted coaxial antenna cables. Rock Mechanics and Rock Engineering. 1989;**22**(1):1-23

[4] Dowding CH, Pierce CE. Use of time domain reflectometry to detect bridge scour and monitor pier movement. In: Proceedings of the Symposium on Time Domain Reflectometry in Environmental, Infrastructure, and Mining Applications; 7-9 September 1994. Evanston, Illinois: U.S. Bureau of Mines; 1994

[5] Osasan KS, Afeni TB. Review of surface mine slope monitoring techniques. Journal of Mining Science. 2010;**46**(2):177-186

[6] Su MB, Chen IH, Liao CH. Using TDR cables and GPS for landslide monitoring in high mountain area. Journal of Geotechnical and Geoenvironmental Engineering, ASCE. 2009;**135**(8):1113-1121

[7] Yin Y, Wang H, Gao Y, Li X. Real-time monitoring and early warning of landslides at relocated Wushan Town, the Three Gorges Reservoir, China. Landslides. 2010;**7**(3):339-349

[8] O'Connor KM, Dowding CH. GeoMeasurements by Pulsing TDR Cables and Probes. Florida: CRC Press, Boca Raton; 1999

[9] Federico A, Popescu M, Elia G, Fidelibus C, Internò G, Murianni A. Prediction of time to slope failure: A general framework. Environment and Earth Science. 2012;**66**(1):245-256

[10] Su MB, Chen YJ. Multiple reflection of metallic time domain reflectometry. Experimental Techniques. 1998;**22**(1):26-29

[11] Drusa M, Bulko R. Rock slide monitoring by using TDR inclinometers. Civil and Environmental Engineering. 2016;**12**(2):137-144

[12] Su MB. Fracture monitoring within concrete structure by time domain reflectometry. Engineering Fracture Mechanics. 1990;**35**(1/2/3):313-320

[13] Su MB, Chen YJ. MTDR monitoring systems for the integrity of infrastructures. Journal of Intelligent Material Systems and Structures. 1999;**10**(3):242-247

[14] Kane WF, Beck TJ. Rapid slope monitoring. Civil Engineering. 1996;**66**(6):56

[15] Su MB, Chen YJ. TDR monitoring for integrity of structural systems. Journal of Infrastructure Systems. 2000;**6**(2):67-72

[16] Dowding CH, Dussud ML, Kane WF, O'Connor KM. Monitor deformation in rock and soil with TDR sensor cables. Geotechnical Instrumentation News. 2003;**21**(2):51-59

[17] Lin YS, Chen IH, Ho SC, Chen JY, Su MB. Applying time domain reflectometry to quantification of slope deformation by shear failure in a landslide. Environment and Earth Science. 2019;**78**(5):123-133

[18] Ho SC, Chen IH, Lin YS, Chen JY, Su MB. Slope deformation monitoring in the Jiufenershan landslide using

time domain reflectometry technology. Landslides. 2019;**16**:1141-1151

[19] Chu YM, Ho SC, Chen IH, Su MB. Applying TDR technology to the automatic monitoring of sliding surface and deformation amount in Lusan landslide area. In: Conference on Sustainable Development and Disaster Prevention in Civil Engineering; June 2019, Kaohsiung, Taiwan; 2019. pp. 635-640 (in Chinese)

[20] Ho SC, Chen IH, Lin YS, Chen JY, Su MB. Using time domain reflectometry for monitoring slope movement in the Jiufenershan landslide. In: 20th Southeast Asian Geotechnical Conference and 3rd AGSSEA Conference, Jakarta, Indonesia; 2018

[21] Chen IH, Lin YS, Ho SC, Chen JY, Su MB. Applying time domain reflectometry technology to measure slope deformation in landslides. In: the 8th Civil Engineering Conference in the Asian region (CECAR8), Tokyo, Japan; 2019. GS-6-2_GS-3, No.2

[22] Shou K, Chen Y, Liu H. Hazard analysis of Li-shan landslide in Taiwan. Geomorphology. 2009;**103**(1):143-153

[23] Shou KJ, Wang CF. Analysis of the Chiufengershan landslide triggered by the 1999 Chi-Chi earthquake in Taiwan. Engineering Geology. 2003;**68**(3-4):237-250

[24] Chang KT, Huang HC. Three-dimensional analysis of a deep-seated landslide in Central Taiwan. Environment and Earth Science. 2015;**74**(2):1379-1390

[25] Lo CM, Feng ZY. Deformation characteristics of slate slopes associated with morphology and creep. Engineering Geology. 2014;**178**:132-154

Chapter 7

The Translatory Wave Model for Landslides

Jónas Elíasson and Þorsteinn Sæmundsson

Abstract

The Saint-Venant equations are usually the basis of numerical models for landslide flows. They are nonstationary and nonlinear. The theory for translatory waves in a prismatic channel and a funneling channel can be used for landslides using the assumption of either turbulent or laminar flow in the slide. The mathematics of translatory waves traveling over dry land or superimposed on another flow are developed. This results in a new slope factor controlling the flow velocity, together with the Chezy coefficient used in previous applications of the translatory wave theory. Flow times for the slide to reach a given destination, slide depth, and velocity can be calculated using the initial magnitude of the flow in the slide. The instabilities of the wave tail are discussed. Three case studies are presented: a submarine slide that started the Tohoku tsunami in Japan, the Morsárjökull rock avalanche in SE Iceland, and the Móafellshyrna slide in central N Iceland.

Keywords: landslide model, translatory wave, landslides, rock avalanche, debris slides, Iceland

1. Introduction

Landslides are a natural phenomenon that more often than not can be described as an environmental disaster. Landslides are responsible for countless human lives, and there is no remedy against their occurrence. The only way to counteract the danger is preparedness, knowing the danger, assessing the hazard, operating a warning system, and keeping people out of the way of the slides, a task all governments and rescue services, civil and military, deal with every day.

Landslides are very different in character, so in the civil protection work, everything matters, area, weather, soil, topography, surface structure, and what triggers the slide. Consequently, the classification system for landslides is very complicated. A very good overview of the modern trend of classification can be found in [1] where types of landslides are discussed separately. A quick overview on what this classification is all about can be found in the course material in [2], where classification systems from 1938, 1958, and 1978 are listed plus a unified classification system.

The classification has to take into account all the danger factors previously mentioned and is therefore quite complicated and in constant development, and a lot has happened since it appeared. For instance, today statistical methods are used to assess the danger [3], and sometimes they bring about scaling laws known from the fractal theory [4]. The statistical methods have not found their way into the classification system.

One of the largest uncertainties and in many instances the very reason for the use of statistics in the science of landslides is the instability problem. Instability can trigger a landslide quite suddenly, and then again, small cracks may be discovered, and they keep widening for a long time; it is quite evident that a landslide will occur at some time, but there is no way to tell when it happens.

Instability problems are well known in soil mechanics, e.g., liquefaction and wave formation in hydrodynamics. Both can be triggered by an external force like earthquake. A lot of research effort is devoted to the problem of landslide triggering [5], sometimes with stochastic input routines [6, 7] and others with models describing the aftermath, for hazard assessment [8], or modeling of the flow of the slide itself [9], sometimes with assistance of remote sensing data [10].

Landslide models can be numerical or analytical. Both are very dependent of the boundary conditions. Both need a physical basis to describe the dynamics of the slide, especially the particle velocity. The simplest, but very efficient, method is to put the velocity proportional to the gradient of the elevation of the surface on the one hand and to the shear force in excess of the shear strength of the soil on the another. An example of the mathematical formulation this assumption leads to is Eqs. 1 and 2 in [5]. The basis of this approach is Newton's law of viscosity.

A necessary condition for this approach to hold is that the inertial terms, other than acceleration, in the full flow equation system can be neglected. This means that all nonlinear terms can be left out. This will be the case for many types of landslides, due to the high viscosity in them. However, many slides will be running faster than that, and they are the most dangerous ones.

In the following we shall discuss the translatory wave landslide model. This model can be used for the front of the landslide, the velocity of it, and how far it flows. The criterion for using the model is that the slide has to be flowing in a regular manner, not just a fall of soil or rock fragment down a steep slope, so many of the types defined in the classification systems and discussions of different types mentioned above will be excluded from this approach. For the others, i.e., regular slides that do actually flow, the translatory theory can in most cases be used.

The translatory model involves analyses that are complicated in nature but lead to a simple system when fully derived. It can be very useful in studies one has to do prior to numerical model studies in defining the area, extent of boundaries, and limits to the various parameters involved. Thus, it can save a lot of trial and error work in constructing the model and elimination of numerical instabilities. And sometimes it produces results that are good enough as demonstrated in the case studies.

2. Theoretical considerations

2.1 Theory of translatory waves

A translatory wave runs down a channel slope with flow that is not constant in time but varies slowly. In treatment of such flow, it is natural to look for waves of fixed shape progressing down a channel. Such waves in shallow water have been investigated earlier; a further treatment is presented in [11] using Chezy's equation for the flow resistance instead of the more complicated Manning's equation. This simplifies the mathematics of the problem. The starting point is the continuity and momentum equations for an arbitrary prismatic channel. For the sake of simplicity, only two cases are treated, that of a wide rectangular inclining flume and a rectangular inclining funnel.

2.2 Translatory waves in a flume

2.2.1 The translatory wave

The equations of Saint-Venant describing flow down an inclining plane can be taken from many textbooks on fluid mechanics or hydrology; they are

$$\text{Equation of continuity:}\ \frac{\delta y}{\delta x}+\frac{\partial q}{\partial x}=0\ q=yv \tag{1}$$

$$\text{Momentum eq.:}\ \frac{\delta v}{\delta t}+v\frac{\partial v}{\partial x}+g\frac{\partial v}{\partial x}\ g(I-I_0)=0 \tag{2}$$

where:
y water depth
q channel flow per unit width
v velocity
x streamwise coordinate
t time
g acceleration of gravity
I friction slope
I_0 channel slope

The system obviously admits a solution where $v = v(x - ct)$ and $y = y(x - ct)$ with c as the constant wave celerity. The momentum equation degenerates to

$$\frac{dy}{d\xi}+\frac{c^2}{C^2gy}=I_0\quad c^2=C^2gIy \tag{3}$$

where C is the nondimensional Chezy constant and $\xi = x - ct$. This equation can readily be integrated.

$$\frac{y}{y_0}+\ln\left(1-\frac{y}{y_0}\right)=I_0\frac{x-ct}{y_0};\quad y_0=0;\quad x-ct\le 0 \tag{4}$$

This is the form of the approaching wave. The wave passes the origin at time zero: $x = t = 0 \Rightarrow y = 0$.

It is easily seen that the wave will continue to run if $I > 0$, i.e., the flow will not stop until the surface has become horizontal. This means that the landslide is of liquefied, or fluidized, soil. This means that the water content of the slide is rather high. This does not have to be the case. The slide's material may very well have finite shear strength. Then it stops at a finite slope.

As time progresses y approaches the normal depth y_0, while the discharge q approaches the value

$$q_0=cy_0=C\sqrt{gI_0}y_0^{3/2} \tag{5}$$

from which the normal depth y_0 and c can be calculated if q_0 is known. Equation (4) can also be used for landslides, as will be demonstrated.

2.2.2 Basic fluid mechanics of landslide flows

Call the shear stress τ and the pressure on the bottom (normal stress) σ in uniform flow of a soil in a landslide down a plane inclined by an angle Θ to the horizontal, and let y' be the distance from the bottom and y the total depth. Then we have from any textbook in hydraulics

$$\tau = \rho g \sin\theta(y - y'); \quad \sigma = \rho g \cos\theta(y - y'); \quad \rightarrow \quad \tau/\sigma = \tan\theta \tag{6}$$

If the shear resistance of the soil is purely frictional with friction angle φ, there is no flow for $\varphi > \Theta$. For cohesive soils this is a little more complicated. Motion starts when $\tau \geq \sigma \tan\varphi + c'$, where c' is the cohesion. In the no flow situation, the cohesion can be a substantial part of the total shear strength of the soil; however, the cohesion is not a material constant for the soil; upon deformation and start of flow, c' is reduced, and after that we can approximate the shear strength of the soil with a dynamical friction angle φ' and a zero cohesion. Then we will have $\varphi' < \Theta$ as condition for flow. This means that the slide will stop when it flows on a slope $\Theta < \varphi'$, and then the shear stress ratio is $\tau_{cr}/\sigma = \tan\varphi'$, which is the critical shear stress ratio. The shear stress in excess of the critical τ_{cr} is then

$$(\tau - \tau_{cr})/\sigma = \tan\theta - \tan\phi'; \quad \rightarrow \quad \tau - \tau_{cr} = \rho g \sin\theta(1 - \tan\phi'/\tan\theta)(y - y') \tag{7}$$

This can be combined with Newton's law of viscosity.

$$(\tau - \tau_{cr})/\rho = g \sin\theta(y - y')(1 - \tan\phi'/\tan\theta) = C_f \frac{dv}{dy'} \tag{8}$$

$$\text{Here we have:} \begin{cases} \text{Laminar flow: } C_f = \text{kinematic viscosity} \\ \text{Turbulent flow: } C_f = \text{eddy viscosity} \end{cases} \tag{9}$$

Kinematic viscosity is normally assumed constant, but eddy viscosity changes with the distance from the wall in accordance with Prandtl's theory of wall turbulence. This makes the difference that the velocity profile will be parabolic in the laminar case and the maximum velocity in a fixed distance in the flow path, i.e., $v_{max} = 1.5\, v_{average}$ for a fixed x in (4). In the turbulent case, v_{max} is closer to $v_{average}$ than that.

Combining this with (4) and (5), we put $c = v_{av}$, and the flow becomes $q = c\, y$. The actual variation of the velocity with y' will for all practical purposes affect the wave form (4) only slightly when the celerity in the landslide is $c << \sqrt{(gy)}$ which is the celerity of the surface gravity wave.

Equations (8) and (9) are similar to those for ordinary channel flow except (6) is multiplied by $(1 - \tan\varphi'/\tan\Theta)$ to get (8), which gives the excess shear stress or the resistance to the flow. Then (9) provides the relation to the velocity through the unknown resistance coefficient C_f. In the laminar case, it is assumed constant, and then we have a flow proportional to the slope, which in a free landscape gives a flow in the direction of the elevation gradient. This is a popular idea for landslide model (see, e.g., [6, 7]).

C_f does not need to be constant, it can depend on the velocity, giving a turbulence-like flow resistance. A simple way to include this possibility is to use Chezy's coefficient defined as

$$c = v_{av} = C\sqrt{(\tau - \tau_{cr})/\rho} = C\sqrt{(gy \sin\theta(1 - \tan\phi'/\tan\theta))} \tag{10}$$

C is not a constant as originally proposed by Chezy but a slowly varying implicit function of the velocity and is thus constant while c is constant. To get an idea about the variation of C with the resistance factor in turbulent flows, one can study Section 15.2 in [12]. There is an implicit relation proposed by ASCE in 1963; it is based on a large volume of experimental data. For low velocity flows, it gives laminar friction. Our implicit relation would have to possess that property too, but otherwise it remains unknown until sufficient experimental data for landslides is available.

Example 1.

We call $S_f = \sin\Theta\ (1 - \tan\varphi'/\tan\Theta)$ the slope factor. It can be found in other models; see, e.g., (5) in [8]. We take the value 0.5. and list in **Table 1** corresponding *C* values for various slides, calculated from (10).

C is proportional with the velocity but varies much more slowly with y. It is therefore a possibility that approximate average regional values for *C* can exist in regions with similar geological settings. *C* for other slope factors S_f can easily be found in **Table 1** by multiplying the *C* values with $a = \sqrt{(0.5/S_f)}$.

2.2.3 Properties of the slide function

When *C* is known, we can turn back to (4) and modify it to fit a slide where the flow resistance is the excess shear stress. In Section 2.3.1, we defined the slope factor for uniform flow from the Chezy coefficient *C* and the bottom slope. Now *C* is defined in the same way but for local surface slope S_f, not bottom slope S_{f0}. With $\xi = x - ct$ as before, the local velocity will be

$$c = C\sqrt{gy \sin\theta \frac{dy}{d\xi}(1 - \tan\phi'/\tan\theta)} = C\sqrt{gyS_f} \tag{11}$$

And (4) is modified to

$$\frac{y}{y_0} + \ln\left(1 - \frac{y}{y_0}\right) = S_{f0}\frac{x - ct}{y_0}; \quad y_0 = 0; \quad x - ct \leq 0 \tag{12}$$

The mathematical properties of Eq. (11) are discussed in [11] and the function plotted in **Figure 1** in [12].

It is obvious from (11) that y approaches y_o very fast with time. According to (4) the flood reaches a depth that is 90% of y_o in $x = 0$ in time $t = 1.4\, y_o/(S_{fo}\, c)$. This is only a few minutes according to the values in **Table 1**; the slower the slide, the steeper is the front of the approaching slide.

The flow q_0 in m^3/s per m width in the slide is found by modifying (5) to (12):

$$q_0 = cy_0 = C\sqrt{gS_{f0}}y_0^{3/2} \tag{13}$$

y/c	**0.5**	**1**	**2**	**3**
1	0.22	0.45	0.89	1.34
2	0.16	0.32	0.63	0.95
3	0.13	0.26	0.52	0.77
4	0.11	0.22	0.45	0.67
5	0.10	0.20	0.40	0.60
6	0.09	0.18	0.37	0.55
7	0.08	0.17	0.34	0.51
8	0.08	0.16	0.32	0.47

Table 1.
C for y in m, c in m/s, and slope factor 0.5.

2.2.4 Change of slope or channel form

If the slope changes, the velocity will adjust to a new depth and velocities in a region just before and just after the slope change as long as the flow is subcritical in the fluid mechanical sense. When the slide flows downhill, the slope will in most cases be diminished. The flow will adapt to the new depth on a short distance.

Equation (5) can be used to find the new depth by using the fact that the flow is the same, upstream and downstream of the slope change. The same approach is used if the channel changes form.

2.2.5 Flow in a funnel-shaped flume

The continuity equation in a channel shaped as a funnel flume is

$$\frac{\partial y}{\partial t} + \frac{1}{r}\frac{\partial (rq)}{\partial r} = 0; \quad q = yv \tag{14}$$

Here we shall only consider the translatory wave when the wave celerity is equal to the water velocity ($v = c$) and the momentum equation degenerates to the form (3). In this case we get a direct solution. But as the flow spreads out, the equilibrium between surface slope and bed slope will come even quicker than before. We can therefore assume the kinematic wave equation to be valid, a few minutes after the front has passed. The kinematic wave routing equation (valid when changes in velocity head are small) reads [11]

$$\frac{\partial Q}{\partial t} + c\frac{\partial Q}{\partial r} = d\frac{\partial^2 Q}{\partial r^2}; \quad d = \frac{q}{2S_{f0}} \tag{15}$$

Using $Q = Q(r - ct)$ as before cancels the left-hand terms, and we get the expression $Q = \alpha\, r\, y\, v$ where α is the opening angle of the funnel.

$$Q = C_1(r - ct) + C_2 \quad or \quad y = (a(r - ct) + b)/r \tag{16}$$

In (15) a and b are unknown constants. They are needed because Eq. 11 cannot fulfill the boundary condition $Q = 0$ when $r - c\,t = 0$. Equation 11 is valid with sufficient accuracy after y has reached 90 % of y_o or more. Many natural channels have a shape

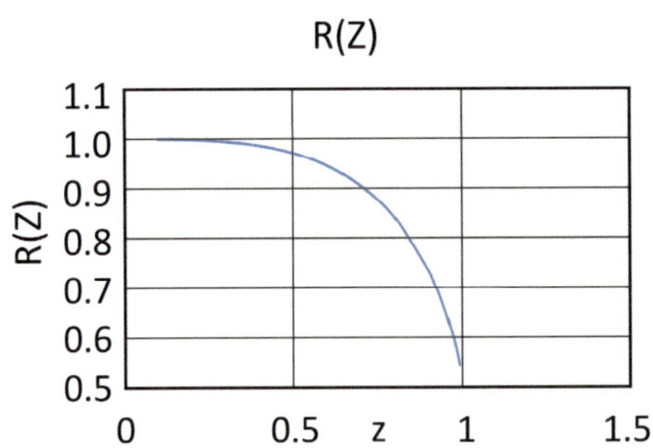

Figure 1.
Chart of the R(Z).

that can be approximated as a series of flumes with constant width and slowly changing funnels. It is therefore quite clear that a translatory wave can flow down any channel, adapt to the changing environment, and still be a translatory wave.

3. Limitations of the theory

3.1 Limitations of the translatory model

The most severe limitation of the translatory model is the very heart of the theory, the progressive wave. The wave starts at some point; for a while there will be either increasing or diminishing flow in this point. While we have increasing or constant flow, we are in the head of the wave. For constant flow (11) and (12) apply, but a flow increase will come as a new translatory wave riding upon the first [11], but (11) will apply, with local y and S_f instead of y_0 and S_{f0}.

3.2 The recessing translatory wave

3.2.1 The recessing wave and the flow head

Diminishing flow is more complicated. Then we are in the tail of the wave head, the wave is recessing, and it is no longer progressing with a constant speed. If it is necessary to calculate this flow, numerical simulation using the full Saint-Venant equations, modified according to Section 2.2.3, may be tried. This involves more advanced programming than for a water waves and has not been tried for landslides using nonlinear friction. In [13] a very advanced model of this kind is presented, using linear friction, but allowing different rheologies, both viscous and non-Newtonian flow, which should give the same possibilities, as nonlinear friction coupled with a slope factor can be used as a tool to imitate non-Newtonian behavior.

The head of the wave is the most damaging part, and the head also determines the length of the damaged area. Therefore, the head flow is the most important, and the flow in the tail is not as interesting in disaster prevention studies.

3.2.2 The laminar wave

If it is clear from the beginning that the velocity is low enough, so all inertial and nonlinear terms can be neglected, and the gradient flow model can be applied in a way like [6, 7]. Using a notation similar to theirs, but using Δy instead of ΔH (with Δ meaning the gradient/divergence operator), we would have a slope factor as in Example 1 and (10).

$$S_f = \sin\theta(1 - \tan\phi'/\tan\theta) = \sin\theta(1 - |\Delta y/(|\tan\theta|)|) \quad (17)$$

Θ is the bottom slope in the direction of the flow. Using the laminar model from (8) and (9), flow equations will be

$$\overline{q} = \overline{v}y = 2/(3C_f)S_f y^3 \Delta y/|\Delta y| \qquad \frac{\partial y}{\partial t} = \Delta\overline{q} \quad (18)$$

$$\frac{\partial y}{\partial t} = \Delta(2/(3C_f)S_f y^3 \Delta y/|\Delta y|) \quad (19)$$

There are several ways to linearize (18), either piecewise or totally, and use it to find the flow and the slide area. This is probably the most popular slide model, but it demands a linear relationship between the flow resistance $(\tau - \tau_{cr})$ in (8) and the surface gradient.

4. The onset of motion

4.1 The initial state

4.1.1 Development of instability

Instability occurs when the shear stress in the soil becomes larger than the critical strength, then a landslide may start. The causes of this can be geological, morphological, physical, or human (see [1] or [2]) or any combination of these causes. A misjudgment of the geological or morphological conditions of the site where a slope is being constructed, or altered, may lead to a landslide. This would be listed as a human cause but can have the simple physical reason that the soil could not withstand the stress increase imposed by the construction work.

Instability of this kind can take time to develop. The soil may grow thicker by time, a bulge can develop due to swelling where the horizontal stress is at its maximum, and a slide will start when the horizontal stress exceeds the passive pressure limit. During the following plastic deformation of the soil, the cohesive strength (c') of the soil is diminished, the plastic deformation is accelerated, the instability has set in, and we have the landslide running possibly a debris flow.

The instability can also set in because c' is reduced by water uptake of the soil, either by infiltration from increased groundwater pressure or soaking of the soil by rain. This reduces the φ' at once and sets in the instability when we have $\varphi' < \Theta$.

A regional study of slopes and lithology, and analysis together with other factors, can bring about important statistics concerning landslide susceptibility and probability of occurrence of landslides caused by the above described instability [3]. This may even be done globally using satellite data [10]. Predicting the timing and location of landslide initiated by developing instability of this kind is very difficult, but considerable success is reported in this field [5, 9].

4.1.2 Slope instability by external causes

External forces can cause slope failure that starts a landslide. These are earthquakes, other tectonic movements, and explosions that rupture the soil and reduce the shear strength of the soil by one quick movement. An instability of this kind depends on the strength of the external force and is even more difficult to predict.

Large earthquakes usually start landslides, both on land and in the sea (submarine slides).

There may be any combination of events that cause landslides. A landslide due to developing instability may be triggered by a small earth tremor. This will cause the landslide to happen sooner than otherwise expected.

5. Practical aspects of the translatory model

5.1 The estimation and use of the Chezy coefficient

The translatory model is based on the Chezy coefficient that can be relatively stable regionally, but this must be established by extensive field investigations.

In field investigations of actual slides, it may be difficult to find the actual investigations begin after the slide stops. Then the total flow of the slide is an important factor. In a point where we have

$$Z = y/y_0 \qquad \text{and} \qquad L = -x/y_0 \tag{20}$$

consider Eq. (11) at time t = 0; the tip of the slide is in x = 0. Let x < 0 be any position, and the relation for the dimensionless total flow past it is

$$Q^* = ZL - R(Z)/S_f; \quad R(Z) = \left(1 + (1 + Z)^2\right) - (1 - Z)\ln(1 - Z) \tag{21}$$

From this we deduce

$$S_f = R(Z)/(ZL - Q^*). \tag{22}$$

The dimensional value of Q is $Q = Q^* y_0^2$. When Q is found from field investigations, (20) is a rather robust formula to find S_f. As seen from **Figure 1**, Z does not vary very much except in the range $0.5\,y_0 < y < 0.9\,y_0$. Outside this range we have 1 or 0.5 for $R(Z)$. If $R(Z)$ is estimated to be 0,5–0,6, we are in the Z range $0.9 < y/y_0 < 1$.

This gives the possibility to use an estimated y_0 value in (20) and find S_f.

Example 1 shows how S_f relates to the hillslope $tan\Theta$ and the critical shear stress ratio $tan\varphi'$ that also relates to the soil parameters.

To find the Chezy coefficient C, there must be some observations of flow times. If the time t_x that it took the slide to move from $-x$ in (19) to $x = 0$, we simply have $c = x/t_x$. Then (12) gives an estimate for C.

But even without an estimate for C, the estimate for y_0 and (11) give us an estimate of how large an area a slide with a known total mass will cover.

5.2 Discussion of the method

As stated in the introduction, the translatory theory can be used for any landslide flowing in a regular manner, which means that the bottom friction and bed slope control the flow. The velocity of vertical falls, turfs, frozen blocks of soil, ice, or snow is not governed by the friction in the flowing phase only at the onset of motion and when the sliding stops. At high velocities a lot of air can be mixed into the flow; this is especially true for avalanches of dry fine-grained snow. In such cases material properties such as density and viscosity may become indeterminable.

For flowing slides, the translatory wave theory can be used for any event, debris flows, or mudflows. In [5–8] slides that this theory can be used for are discussed. But it is also possible to apply it for unusual cases as submarine slides and rock avalanches as will be demonstrated in the next chapter.

The main application of the translatory theory is to use the information available of source area material properties and magnitudes, rundown slopes, and flow depths to estimate the flow coefficients S_f and C. A good database for these coefficients will make it possible to estimate them at sites where a slide danger has been identified and the slide is expected to fall anytime.

In such cases a danger area has to be estimated for use by the civil protection authorities. They can declare the area off limits (all visits forbidden), close roads, and order areas to be evacuated. It is especially in the case of evacuations and closing roads for all traffic that an overcautious and too large a danger area estimate can come under severe criticism.

When the size of a danger area is estimated, there has to be a safety margin. In construction it is decided by the safety margin to put on material strength and design loads. Using S_f and C to find a danger area, the safety margin can be decided in the same way. The most likely values for S_f and C have to be found first and the safe values to be used in the estimation decided afterwards.

6. Case studies

6.1 Case study 1: the Tohoku submarine slide

6.1.1 Location and event

The most famous tsunami event of recent years hit Honshu, the big island in Japan, on Friday 11 March 2011 at 05:46 UTC. Its location was off the Pacific coast of Tohoku, Japan. The earthquake event and the devastation caused by this tsunami is very well documented; here we shall follow [14].

The cause was a M_w 9.0 (magnitude moment) undersea megathrust earthquake. The epicenter was approximately 70 km east of the Oshika Peninsula of Tohoku with the hypocenter in approximately 30 km deep water.

As most earthquakes above M_w 7.0 do, it caused a tsunami. The tsunami was much bigger than what could be expected and caused a massive devastation. Tens of thousands of people suffered loss of life or all their possessions. Houses and other structures were swept away from large areas. The town of Sendai was very badly hit and 13 other Japanese cities.

The most serious incident, however, was a meltdown in the Fukushima Daiichi Nuclear Power Plant, when the tsunami proved too high for the coastal defenses and flooded the emergency generators.

6.1.2 Properties and data

In [14] it is suggested that the bottom deformations that caused the very strong Tohoku tsunami in the Pacific Ocean can be explained by a submarine landslide. This suggests that the coseismic slip of the earthquake triggers a sliding of the surface sediments. In combination they cause the bottom deformation.

Finally, it can thus be concluded that the coseismic slip and the landslide are both responsible for the Tohoku tsunami in March 2011, not the coseismic slip alone.

From the data obtained in the exploratory drilling at the site shown in Figure 2 in Ref. [14], it was concluded that the tsunami was caused by the mass movement shown in Figure 3 in Ref. [14]. The slide triggered by the earthquake amplifies the tsunami.

This landslide is not very high compared to what has happened elsewhere. But it is quite possible that the coastal defenses of Fukushima would have saved the emergency generators and prevented the nuclear disaster had the landslide not happened.

The main data for the slide were originally deduced from [15–17]. In [14] we find the following:

Bottom slope is 9 %, c = 2 m/s, and y_0 = 8 m. The length of the slide is 117.5 km, the width of the slide is 110 km, but the length of the flow path is only 60 m, and the flow time is thus 30 s. The volume moved by the slide is 96.8 km^3.

6.1.3 Evaluation

Using **Table 1** and the bottom slope, we find a C value of 0.32. Corrected for $S_f \approx 0.1$, we get the value of the Chezy coefficient C = 0.71.

This shows that in assessing the tsunami risk in the Pacific coastal regions of Japan, the landslide risk must be considered. This fact may result in that considerably larger events than the Tohoku tsunami are possible if a larger slide than this 8-m-thick slide is released.

The assessment of this possibility of larger slides can be difficult; there is considerable uncertainty in estimating the height (y_0) of possible landslide, but this may be done using drilled exploration wells, like the one shown in Figure 2 in Ref. [14]. The flow time will be the estimated duration of the earthquake, so using the C found here, there may be a possibility to estimate the volume moved by a slide using (20)–(21) and thereby the strength of the corresponding tsunami.

6.2 Case study 2: the rock avalanche on the Morsárjökull outlet glacier, SE Iceland

6.2.1 Location and event

On March 20, 2007, a large rock avalanche fell on the Morsárjökull outlet glacier, in the southern part of the Vatnajökull ice cap, in Southeast Iceland (**Figure 2**). This rock avalanche is considered to be one of the largest rock avalanches which have occurred in Iceland during the last decades [18].

The Morsárjökull glacier is a small outlet glacier on the southern side of the Vatnajökull ice cap in Southeastern Iceland, located between the Þorsteinshöfði Mountain (1343 m) and Skarðatindur Mountain (1385 m) to the east and the Austurhnúta peak (1084 m) in the Miðfell Mountain (1430 m) to the west (**Figure 2**). The glacier is about 4 km long and 750–900 m wide in the uppermost part and around 800–900 wide in the lower part. Its surface is around 3.3 km^2. The glacier is surrounded by steep mountain slopes and is categorized as an icefall glacier with two ice falls. In the uppermost part of the glacier, below the ice falls, the glacier surface is steep, but below the icefalls the surface inclination is around 85/

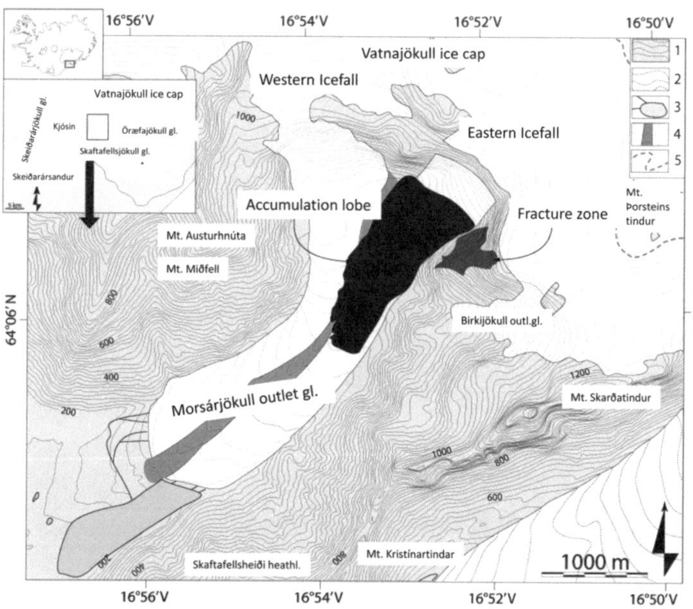

Figure 2.
Location map of the Morsárjökull outlet glacier in SE Iceland. The fracture zone and the accumulation lobe marked with gray color. Legend: (1) 20 m contour lines on land, (2) 20 m contour lines on glacier, (3) proglacial lake, (4) medial moraine, and (5) equilibrium line (ELA).

1000 m. A proglacial glacier lake occurs in front of the eastern side of the glacier snout. The lake, which is today around 1.8 km long and 350–400 m wide, begun to form around 1945. This lake has increased considerably in size during the last decades. About 1.6 to 1.8 km from the glacier snout, a system of end moraines occurs crossing the valley floor, marking the maximum extent of the glacier around 1910. The glacier is drained by the Morsá Glacier River, which merged into the old Skeiðará River gorge approximately 12–13 km down glacier.

It is believed that the rock avalanche fell in two separate stages; the main part fell on March 20 and the second and smaller one on April 17, 2007. In both instances surface tremors were detected on seismographs, and on April 17 workers at the Skeiðarársandur sandur plain heard a rumbling noise from the glacier.

The rock avalanche was initiated from the north face of the headwall above the uppermost part of the glacier (**Figure 3**). The scar in the rock-face is around 330 m high, reaching from about 620 m up to 950 m and about 480 m wide on average, showing that the main part of the headwall collapsed. From a digital elevation model of the headwall, based on aerial photographs from 2003 and 2007, the total volume of debris is estimated to be around 4 to 4.5 million m^3 or about 10–12 million tons. The rock avalanche debris fell first straight onto the glacier and then turned 90° downward. No indication of any deformation or erosion of the glacier surface was observed, which is similar to the description of the rock avalanche which fell on the Sherman glacier in Alaska in 1964 [19, 20].

6.2.2 Properties and data

The accumulation lobe is on average 1,5 km long, with a maximum length of 1.6 km, reaching from 520 m a.s.l. down to about 350 m a.s.l. Its width is from 125 m to 650 m or on average 480 m. The total area which the lobe covers is about 720.000 m^2 which was approximately 1/5 of the glacier surface in 2007. The thickness of the debris lobe was measured more than 8 m in places, but its mean thickness was somewhere between 5.5 and 6.2 m. In 2007 the debris mass is coarse-grained and boulder-rich with little fine material. Blocks over 5 to 8 m in diameter are common on the edges of the lobe up to 1.6 km from the source. The largest block which was observed in the accumulation lobe is around 800 tons located on the western side of

Figure 3.
The headwall above the eastern icefall in the uppermost part of the Morsárjökull outlet glacier and the accumulation lobe (Photo Þorsteinn Sæmundsson 2007).

the debris lobe, approximately 800 m from the scar. On the surface of the accumulation lobe flow line structures and longitudinal ridges, up to 3–4 m high were observed. Similar phenomena have been described from other rock avalanches on glaciers, e.g., the Sherman glacier in Alaska [20] and the Little Tahoma Peak glacier in the Washington state [21].

It is evident that the Morsárjökull glacier has retreated considerably during the last century, and during the last decade, the melting has been very rapid. Therefore, it is thought that undercutting of the mountain slope by glacial erosion and the retreat of the glacier are the main contributing factors leading to the rock avalanche. The glacial erosion has destabilized the slope, which is chiefly composed of palagonite and dolerite rocks, affected by geothermal alteration. Hence a subsequent fracture formation has weakened the bedrock. The exact triggering factor, however, is not known. No seismic activity or meteorological signal, which could be interpreted as triggering factors, such as heavy rainfall or intensive snowmelt, was recorded prior to the rock avalanche.

6.2.3 Evaluation

A quick calculation using (11) sh0ws that there is very little difference between the average thickness and y_0. The upper section is around 500 m long, making the lower section 1000 m. The glacial surface just below where the slide stops inclines by 5.6 %; this is put as $tan\phi$'.

It is necessary to estimate the flow time of the slide in order to find the Chezy coefficient C; this is found from a tremor recorded in a nearby earthquake station to be 100 seconds, but this estimate is very uncertain. However, it affects only c and C. **Table 2** shows the data and the calculation of the main parameters of the slide.

The main results show similar values for the upper and lower C's. Note the large difference from the underwater slide in Case 1. Had the C values been known, the time could have been computed.

	Upper section		**Lower section**	
	Source	**Value**	**Value**	**Dim**
y_0	Data	5	7	m
Θ	$Arctan(S_{fo})$	9.0	5.1	Deg
S_{fo}	Data	0.16	0.09	NA
S_f	$sin\Theta\ (1 - tan\varphi'/tan\Theta))$	0.10	0.03	NA
Time	100 totals	25	75	sec
Length	1500 total	500	1000	m
c	Length/time	20	13.3	m/s
C	$C = c/\ (g\ y_0\ S_f)$	8.8	8.7	NA

Table 2.
Upper section of the Morsárjökull slide.

6.3 Case study 3: the debris slide in the Móafellshyrna Mountain in the Tröllaskagi peninsula, central north Iceland

6.3.1 Location and event

On September 20, 2012, a large debris slide occurred at the northwest facing slope of the Móafellshyrna Mountain in the Fljótin area situated in the Tröllaskagi peninsula, an area in central north Iceland. The local residents of the farm Þrasastaðir witnessed the release of the landslide. The area was heavily eroded during the last glaciation period Iceland. **Figure 4**, modified from [22], shows the site, the location, and an aerial picture of the extent of the slide. The onset of the slide is traced to the special geology of the mountain (**Figure 4**) and thawing of permafrost in the sediments at an approximate elevation of 900 m (meters elevation above sea level) at the top of the Móafellshyrna Mountain. The thawing was proven to be the primary reason for the weakening of the sediments in the source area, but the triggering of the event was a combination of heavy rains and three earthquakes of Richter magnitude 4.2–4.5 on September 19 and 20.

6.3.2 Properties and data

The slide originates (at the headscarp) at an elevation of 880 m and runs all the way down to 330 m (**Figures 4** and **5**). The total debris volume is estimated to be 312,000–480,000 m^3 [22].

There are three particularly interesting problems associated with this slide. Firstly, in it are large blocks of frozen sedimentary rocks; the biggest one is 12 meters wide and 4 meters high (**Figure 6**). This large block is an integral part of the debris flow, it is carried almost halfway down the mountain, and it stops at the elevation 495–505 m and has carved an erosion channel into the mountainside, shown as black streak in the talus (rock mass at the foot of a cliff) slope in **Figure 5**, but it stops at around 500 m a.s.l.

Secondly, the massive frozen blocks inside the main debris flow will affect the flow velocity and the velocity distribution within the flow. The erosion channel

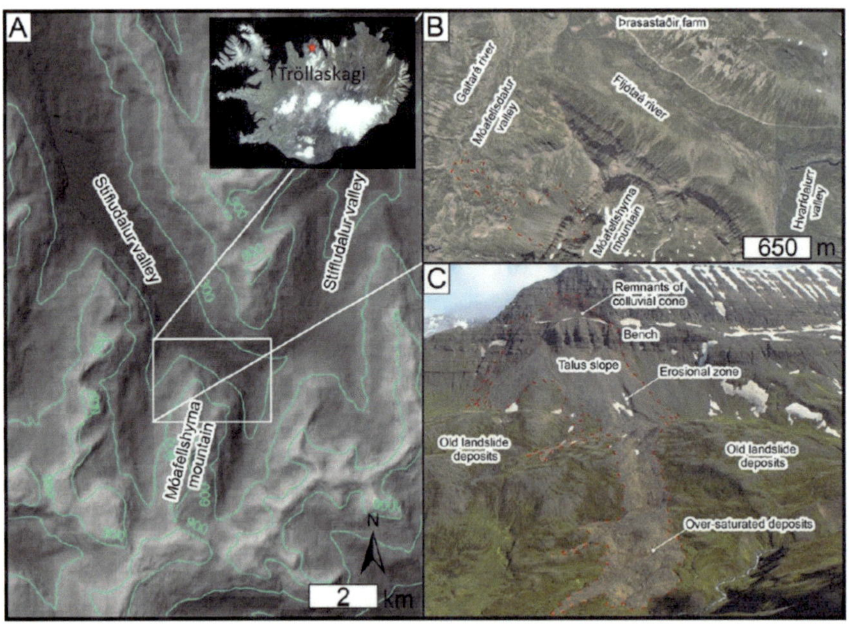

Figure 4.
The geographic setting of the Móafellshyrna site, Tröllaskagi peninsula, central northern Iceland. (A) Hillshaded digital elevation model and contours (in green, meters above sea level) of the Móafellshyrna region. Elevation data are from EU-DEM from GMES RDA. (B) Aerial photograph of the Móafellshyrna site taken before the slide in 2012 from the website Samsyn. (C) Oblique photo of the Móafellshyrna debris slide in July 2015 taken by Costanza Morino. In panels B and C, the perimeter of the slide is marked with red dashed line taken from the trimble data for the deposits and reconstructed from photographs for the upper part; in C this line has been manually traced onto the oblique image. Modified from [22].

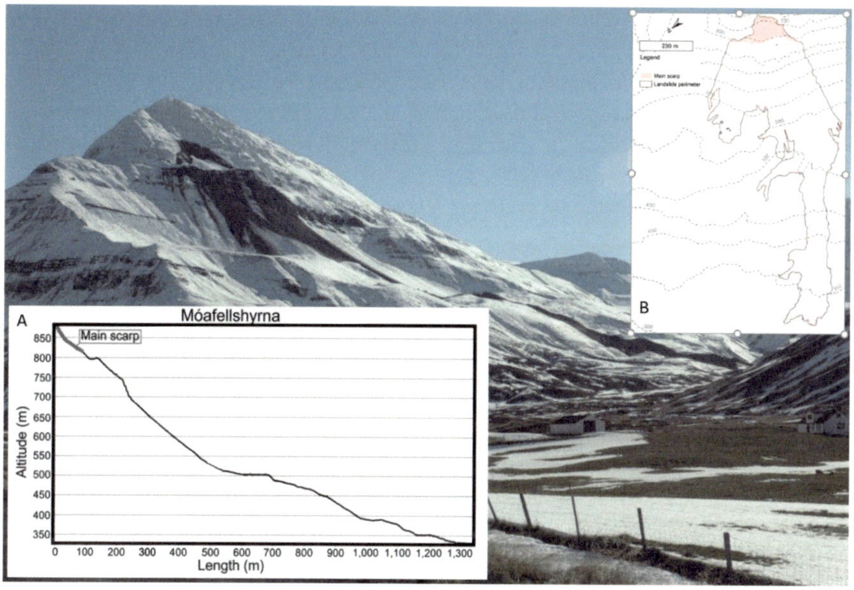

Figure 5.
The Móafellshyrna landslide (photo taken few hours after the slide on September 20, 2012). (A) The topographical profile of the Móafellshyrna landslide. (B) Distribution map of the landslide (photo by Þorsteinn Sæmundsson 2012). Modified from [22, 23].

Figure 6.
Landslide debris and an ice-cemented block resting at 495–505 m a.s.l. The block is around 12 m wide and 4 m high (photo by Þorsteinn Sæmundsson September 29, 2012).

shows that the part of the slide flowing in it has lower velocity than the rest of the flow, which means that the main flow is providing a part of its momentum to the blocks. This may result in a thicker slide that does not run as long out as a slide of uniform material would have done.

Thirdly, the blocks run over a 100-meter-high cliff below the source area at 690. Here they pick up momentum and carve the channel before they stop, 170 m higher up than the rest of the slide which runs about 700 meters farther out (**Figure 5**). The force of the flow on the blocks is unable to transport them any further in spite of the momentum gain. This acts as flow resistance on the part of the slide flowing by and may be a significant factor in determining the flow properties down streams, e.g., that it stops where it does and does not flow any further.

Figure 5A and **B** explains the properties of this very complicated slide visually.

The following description of the slide shows where erosion and deposition occurred in the slope (**Figure 5**) [22, 23].

- The debris slide was detached from a main scarp at an elevation of 880 m a.s.l. The source material of the landslide was a colluvial cone. Part of the cones front, slides off a steep 100 m high rockwall, at around 800 m a.s.l. It is estimated that the volume of deposit which broke the frontal part of the colluvial cone was around 120,000–180,000 m^3. The headscarp has an average slope of around 30–35°.

- Below the rockwall at around 700 m a.s.l., a 300-m-long, 80–100-m-wide, and 8–10-m-deep channel was carved into the 200-m-high talus slope deposits, entraining around 192,000–300,000 m^3 of material. The average slope of the talus is around 30°.

- At 495–505 m a.s.l. at the foot of the talus slope, some of the landslide deposits came to a rest, where the average slope angle is less than 10°.

- The rest of the landslide material traversed down the mountain slope from 495 m and finally came to a rest at 330 m. The uppermost 100 m of the slope are steeper with an averaged slope angle of around 20–25° and in the lowermost 70 m of the slope around 10–15° angle.

6.3.3 Evaluation

The slide may be translatory from the resting position of the big block at around 500 m a.s.l. until it stops. The section around it is then the starting point of the downstream flow, and its initial conditions are the time history of the velocity and flow depth in that section.

From **Figure 5** we see that the slide stops at the bottom slope of $tan\varphi' = 50/400 = 0,125$ or $\varphi' = 7°$. The width of the slide can be approximately estimated from **Figures 4** and **5**. According to Section 6.2.2, the total flow mass to run from the talus slope from 700 to approximately 500 m a.s.l. is estimated to be 192,000–300,000 m^3. This provides an estimate that the flow channels from 500 down to 330 m has an average depth of 6 m. If the flow channels are supposed to have a parabolic shape, this corresponds to 2/3 of their maximum depth. The profiles and location of these channels are shown in **Figures 5A** and **B**. By assuming the same Chezy coefficients as in **Table 2**, we can calculate **Table 3**.

The results indicate that the translatory theory can model the slide quite well. The flow velocity c is found and used to calculate the flow times from 505 to 400 m height. It takes the flow 25,5 s to reach the 330 m elevation but only 8,5 s to run from 495 to 400 m. However, it takes the total mass 28 seconds (total flow 28) to flow from 495 to 400 m. This is longer time than the run from 400 to 330 m, so some material will be left between 495 and 400 m when the downstream end stops, making some accumulation to take at that height interval.

This is an example of computing slide properties from known C values in a fast and effective manner. The velocity c in the steeper part of the lower part of the slide from 495 to 400 m is high, certainly high enough to produce the rumbling noise

	495–400 m		400–330 m	
	Source	**Value**	**Value**	**Dim**
y_0	2/3 of the max. value	6	6	m
Width	**Figure 5B**	80	130	m
Θ	$Arctan(S_{fo})$	19	10	Deg
S_{fo}	**Figure 5A**	0,33	0,17	NA
S_f	$sin\Theta\ (1 - tan\varphi'/tan\Theta))$	0,34	0,17	NA
Time	Length/*c*. Total flow 28	8,5	25,5	sec
Length	**Figure 5A**	263	360	m
c	$c= C\ (g\, y_0\, S_f)$	31	14	m/s
C	**Table 2**	8,8	8,7	NA

Table 3.
Two sections in the Móafellshyrna slide.

heard in the nearby farm. The velocity from 400 to 330 m is still high; it takes a slope of 70 meters to stop it.

7. Conclusion

The translatory wave theory is an old tool from the science of hydrodynamics that can be adapted to landslides. However, it is not possible to say that it may be used for certain classes or groups of landslides in the geological classification system of landslides. The theory can be used for all slides that flow downhill with a flow governed by gravitational forces, where the slide has a wavelike front and the slide follows as a frontal wave with the same velocity.

The translatory wave equation can be modified to describe this form of a landslide by introducing a new parameter, the slope factor S_f into the flow equations as to control the excess shear force in the landslide and its relation to the actual shear strength of the flowing mass. The slope factor controls how great a part of the actual shear force is in excess of the shear strength and thereby the velocity of the flow.

The slope factor has to be assumed piecewise constant, and the flow stops when it becomes zero. This model can give a very good picture of the flow and can save valuable time and effort, when a more detailed numerical treatment is necessary.

The translatory model is used for moving slides; it cannot predict the onset of a slide. But when in motion, it can predict the length and thickness of the slide from the knowledge of the ground slope and the soil parameters, the Chezy coefficient, and the slope factor. And vice versa, for a mapped slide, it can predict the soil parameters.

There are three studied cases presented, first the famous Tohoku tsunami in Japan, which became very strong because of the earthquake that started the tsunami and a landslide that amplified the tsunami. Both Chezy's coefficient and the slope factor are found.

The second case study is a landslide down the Morsárjökull outlet glacier in Iceland, starting as a rock slide, but it turns downhill into a glacier and turns into a flow slide, in clear contrast to the Tohoku submarine slide. This slide flows first down a steep upper slope and then down a milder lower slope and then finally stops on an even milder slope with a flow factor zero. The results for the upper and lower sections show different slope factors, the lower one belonging to the lower section. The lower thickness value belongs to the upper section, but the Chezy coefficients are similar for the two sections. There is reason to believe that further research can uncover a pattern for the Chezy coefficient for similar sites. Then runout lengths and risk area can be approximately predicted for slides of known volume, known from developing cracks or other discernable geological evidence.

The third case study is a landslide down the Móafellshyrna Mountain in North Iceland. It has previously been shown that this slide can be modeled with the RAMS model. Here it is demonstrated that it can also be modeled using the Chezy coefficients estimated in case study 2. The possibility to keep a data set containing Chezy coefficients that can be used in future estimates is a considerable help in hazard assessments.

Author details

Jónas Elíasson* and Þorsteinn Sæmundsson
Faculty of Engineering and Natural Sciences, University of Iceland, Reykjavík, Iceland

*Address all correspondence to: jonase@hi.is

References

[1] Highland, LM, Bobrowsky P. The Landslide Handbook—A Guide to Understanding Landslides, United States Geological Survey, Landslide Program and National Landslide Information Center, Mail Stop 966, Box 25046, Denver Federal Center, Denver, Colorado, 80225 USA. 2008. Available from: https://pubs.usgs.gov/circ/1325/

[2] van Westen C. Introduction to Landslides, Part 1: Types and Causes. Available from: http://www.adpc.net/casita/Course%20Modules/Landslide%20hazard%20assessment/Landslides_types_and_causes.pdf

[3] Komac M. A landslide susceptibility model using the analytical hierarchy process method and multivariate statistics in perialpine Slovenia. Geomorphology. 2006;**74**(1-4):17-28

[4] Guzzetti F, Malamud BD, Turcotte DL, Reichenbach P. Power-law correlations of landslide areas in central Italy. Earth and Planetary Science Letters. 2002;**195**(3-4):169-183

[5] Casadei M, Dietrich WE, Miller NL. Testing a model for predicting the timing and location of shallow landslide initiation in soil-mantled landscapes. Earth Surface Processes and Landforms: The Journal of the British Geomorphological Research Group. 2003;**28**(9):925-950

[6] Crozier MJ. Prediction of rainfall-triggered landslides: A test of the antecedent water status model. Earth Surface Processes and Landforms. 1999; **24**(9):825-833

[7] Hergarten S, Neugebauer HJ. Self-organized criticality in a landslide model. Geophysical Research Letters. 1998;**25**(6):801-804

[8] Borga M, Dalla Fontana G, Da Ros D, Marchi L. Shallow landslide hazard assessment using a physically based model and digital elevation data. Environmental Geology. 1998;**35**(2-3): 81-88

[9] Tsai TL, Yang JC. Modeling of rainfall-triggered shallow landslide. Environmental Geology. 2006;**50**(4): 525-534

[10] Farahmand A, AghaKouchak A. A satellite-based global landslide model. Natural Hazards and Earth System Sciences. 2013;**13**(5):1259-1267

[11] Elíasson J, Kjaran SP, Holm SL, Gudmundsson MT, Larsen G. Large hazardous floods as translatory waves. Environmental Modelling and Software. 2007 Oct 1;**22**(10):1392-1399

[12] Douglas JF, Gasiorek JM, Swaffield JA. Fluid Mechanics. 3rd Edn. England: Longman Scientific & Technical, Essex CM20 2JE; 1995. p. 819

[13] McDougall S, Hungr O. A model for the analysis of rapid landslide motion across three-dimensional terrain. Canadian Geotechnical Journal. 2004; **41**(6):1084-1097

[14] Eliasson J. Earthquake-generated landslides and tsunamis. In: Earthquakes-Impact, Community Vulnerability and Resilience. Rijeka: IntechOpen; 2019

[15] JAMSTEC. Causal Mechanisms of Large Slip during the Tohoku Earthquake of 2011. Revealed through Hydraulic Analysis of Fault Drilling Samples from the Deep–Sea Scientific Drilling Vessel Chikyu; Japan Agency for Marine-Earth Science and Technology. 2013. Available from: http://www.jamstec.go.jp/e/about/press_release/20131008/

[16] Land and Sea Areas of Crustal Movement and Slip Distribution Model of the Tohoku-Pacific Ocean

Earthquake: Coseismic Slip Distribution Model. Report by the Geographical Survey Institute, Japan. 2011. Available from: http://www.gsi.go.jp/cais/topic 110520-index-e.html

[17] Fujii Y, Satake K, Sakai SI, Shinohara M, Kanazawa T. Tsunami Earthquakes - Impact, Community Vulnerability and Resilience source of the 2011 off the Pacific coast of Tohoku earthquake. Earth, Planets and Space. 2011;**63**(7):55

[18] Sæmundsson Þ, Sigurðsson IA, Pétursson HG, Jónsson HP, Decaulne A, Roberts MJ, et al. Bergflóðið sem féll á Morsárjökul 20. mars 2007 –hverjar hafa afleiðingar þess orðið? Náttúrufræðingurinn. 2011;**81**(3–4): 131-141 (in Icelandic)

[19] Post AS. Alaskan Glaciers: recent observations in respect to the earthquake-advance theory. Science. 1965;**148**:366-368

[20] Shreve RL. Sherman landslide, Alaska. Science. 1966;**154**(3757): 1639-1643

[21] Fahnestock RK. Little Tahoma Peak rockfalls and avalanches, Mount Rainier, Washington, USA. In: Voight B, editor. Developments in Geotechnical Engineering. Vol. 14. Part A. New York: Elsevier; 1978. pp. 181-196

[22] Sæmundsson Þ, Morino C, Helgason JK, Conway SJ, Pétursson HG, Barceló D, et al. The triggering factors of the Móafellshyrna debris slide in northern Iceland: Intense precipitation, earthquake activity and thawing of mountain permafrost. In: Barceló D, Gan J, editors. Science of the Total Environment. 2018;**621**:1163-1175

[23] Morino C, Conway SJ, Sæmundsson Þ, Helgason JK, Hiller J, Butcher FEG, et al. Molards as an indicator of permafrost degradation and landslide processes. Earth and Planetary Science Letters. 2019;**516**:136-147

Chapter 8

Techniques to Evaluate and Remediate the Slope Stability in Overconsolidated Clay

Herman Peiffer

Abstract

In this chapter, the origin and remediation of an important sliding in the overconsolidated Boom Clay in Kruibeke (Belgium) is discussed. Local and environmental factors caused an unstable slope about 30 m deep, A larger sensitivity to erosion resulted finally in the instability of the slope. Because of the formation of fine cracks in the soil there was a possibility for the water to penetrate in the clay close to the surface, resulting in the presence of higher water pressures. Also, the presence of the excavator on top of the slope during exploitation had an important impact on the stress state of the soil. Both an analytical and numerical approach were used to estimate the factor of safety. Because of the change of the soil characteristics, the factor of safety decreases, which can be estimated through a numerical analysis (using the Strength Reduction Method). This chapter also discussed the applied techniques for the remediation using numerical analysis. Also, the importance of the field test is discussed. An integrated approach, using numerical analysis and field tests in combination, is capable of predicting the instability. This approach can also be used to evaluate the stability of the slope after remediation.

Keywords: slope stability, monitoring, in situ testing, DMT (dilatometer), rehabilitation

1. Introduction

A natural or an artificial slope in a clay deposit results in a change of the morphological properties and the stress state in the soil This evolution of the stress state can be measured by in situ tests considering the variations over time of the lateral stress at rest K_0 [1].

K_0 is an important parameter in the design and monitoring in many geotechnical applications (retaining walls, building foundations, and definitely the stability of slopes). For the evaluation of the stability of slopes, it can be estimated out of the results of field tests (Pressuremeter Tests [1], Vane Tests and Piezometer Cone Tests [2], Dilatometer Test [3–5] or numerical modeling, usually considering elasto-viscoplastic models.

Although the analysis of the lateral stress, in case of slope stability, also can be made based on the results of laboratory tests [6–8], one of the main problems is that K_0, due to the natural variations of horizontal and vertical stresses, at each location depends on the actual stress conditions.

Silvestri [9] performed a post slide analysis using the critical stress curve method for two failures that occurred in overconsolidated sensitive clay. The results of the analysis yielded the shear strength parameters. They found that the strength at the time of failure was due to the fissured nature of the clay, the spreading of the discontinuities caused by stress relief after excavation, and subsequent desiccation of the exposed clay mass.

Grefsheim [10] made a back analysis calculation from full scale landslides in overconsolidated clays using stress-strain plots based on the results of laboratory tests. The author compared the results of his research with those for similar overconsolidated clay soils available in the literature (London Clay and Panama Canal Clay). The correlation was made based on the relationship between shear strength and the plasticity index.

Besides the stress state of the soil, also the groundwater conditions and groundwater pressure plays an important role in the stability of slopes in clay.

The groundwater pressure also determines the stress state in the soil and plays an important role in the equilibrium of the slope.

Chandler [11] discussed twelve case records of failures of slopes in the Upper Lias Clay cutting slopes. The author evaluated that the groundwater behavior based on the results of piezometric measurements. The results of effective stress stability analyses revealed a similar pattern to that found earlier in London Clay (which has the same geological origin as the Boom Clay).

Dehandschutter [12] studied the effect of fracture-development in the Boom Clay on the fluid-flow changes (fracture permeability) and mechanical strength of the soil. This fracture-development is partially due to the soil relaxation after excavating the pit.

D'Elia et al. [13] (1996) and Totani et al. [2] used the Dilatometer Test for the evaluation of the soil conditions after a sliding occurred.

Slope stability is influenced by many factors, among which are subsoil structure, hydrologic conditions (e.g., soil moisture, groundwater position), climatic factors (e.g., rainfall), and the hydraulic boundaries. This chapter presents a multi-disciplinary approach for the remediation of the stability of slopes, which includes a numerical analysis based on the results of the in situ geotechnical monitoring, laboratory geotechnical tests (for the determination of the shear resistance and stiffness of the soil) and piezometric measurements. Thus, a reliable 2D model of the subsoil can be obtained, with well-defined boundaries on which it is possible to apply appropriate hydraulic conditions. This geotechnical model has been used for studying the remediation of the stability of the slope.

2. Methodology

2.1 General

In this chapter, a comparative analysis is made considering the results of in-situ tests in order to evaluate the sensitivity of these testing methods for the evaluation of slope stability.

The research was conducted in Kruibeke (Belgium) (**Figure 1**.). This research site has overconsolidated Boom clay. In the period of 1963–2010, 30 meters deep pit was excavated in the Boom Clay. The clay was used for the fabrication of expanded clay granulates.

In 2010, the borders of the concession were reached, and the exploitation of this pit was stopped. One year later, in 2011, a first (limited) instability of the slope of this pit occurred (sliding). Apparently, it became clear that more stability problems

Figure 1.
Layout of the slide (2014) (as adapted from Peiffer (2016) [5]).

could be expected. In 2012 and 2013, additional (limited) slidings occurred. In the beginning of 2014, it was decided to set up an extensive monitoring program in order to evaluate the risk of further instability of the slopes.

In addition to the monitoring of the settlements and the pore water pressures in the environment, in-situ-tests were carried out on a regular basis to evaluate the stability of the slope.

For the analysis, cone penetration tests (CPT) and dilatometer tests (DMT) were executed. A layout of the test locations is presented in **Figure 2a**. Four campaigns have been organized:

- January 2014—DMT 1 to 5 (before the major sliding) (**Figure 2a**)
- July 2014—DMT A to F (about 1 month after the sliding) (**Figure 2a**)
- January 2015—DMT I to IV (after stabilisation) : only DMT 1 was conducted in the neighborhood of the sliding (**Figure 2a**)
- March 2016—DMT1' (in the immediate neighborhood of DMTI)

2.2 Site investigation: CPT-tests and DMT-tests

2.2.1 Cone penetration test (CPT)

The Cone Penetration Test (CPT) is internationally one of the most widely used and accepted test methods used to determine geotechnical properties of soils and the soil strategy.

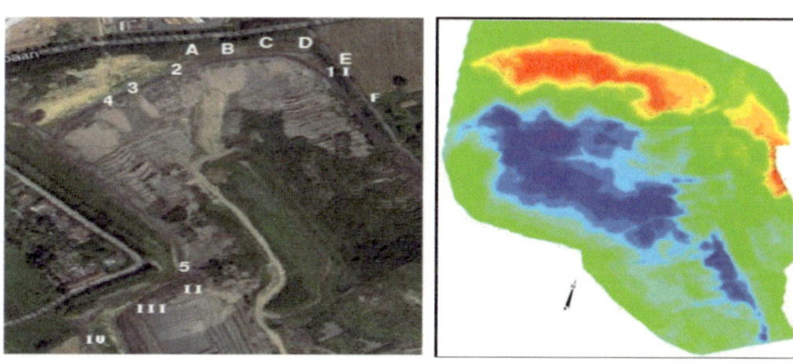

vertical displacement slope			
Number	Min difference	Max difference	Color
1	−9.000	−8.000	
2	−8.000	−7.000	
3	−7.000	−6.000	
4	−6.000	−5.000	
5	−5.000	−4.000	
6	−4.000	−3.000	
7	−3.000	−2.000	
8	−2.000	−1.000	
9	−1.000	0.000	
10	0.000	1.000	
11	1.000	2.000	
12	2.000	3.000	
13	3.000	4.000	
14	4.000	5.000	
15	5.000	6.000	
16	6.000	7.000	
17	7.000	8.000	
18	8.000	9.000	

Figure 2.
Photo view of the site with indication of the DMT test locations (a) and presentation of the magnitude of the vertical movement (b) (as adapted from Peiffer (2016) [5]).

The test method consists of pushing an instrumented cone, with the tip facing down, into the ground at a controlled rate of 2 cm/s. The typical cone tip has a cross-sectional area of 10 cm^2, corresponding to a diameter of 36 mm (**Figure 3**).

During the CPT-test, the cone resistance q_c is measured over the whole depth of the soil profile, generally up to more than 20 m below the surface or until the cone reaches a hard layer. A typical setup for the execution of a CPT-test is presented in **Figure 4**. The cone is pushed in the soil using a counterweight (in general, a heavy truck) that is positioned horizontally. The cone resistance is recorded using force sensors in the tip. These values allow for a good understanding of soil behavior. In particular, the strength of the soil and the relative density (compaction degree) can be determined.

Figure 3.
The CPT-cone.

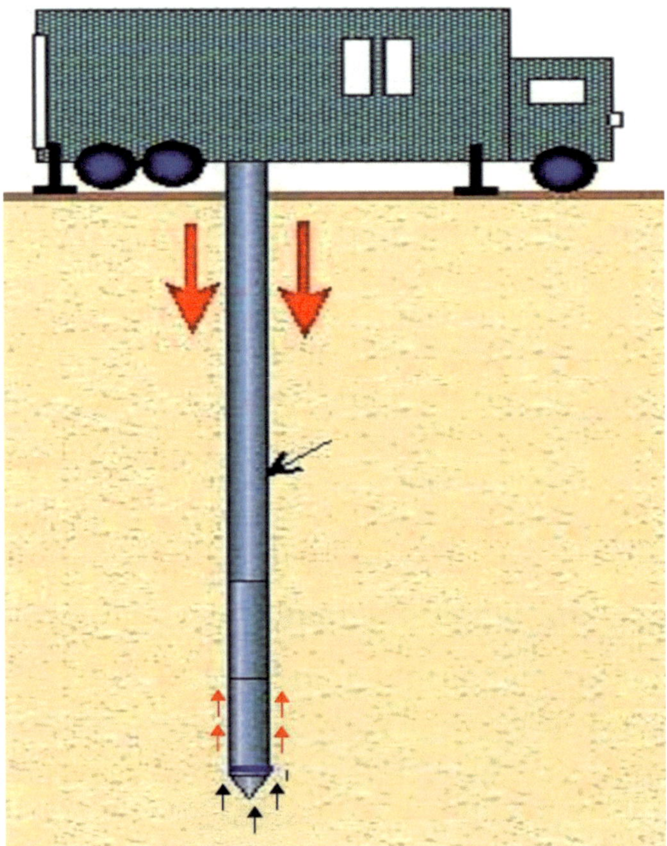

Figure 4.
Execution of the CPT-test.

A typical application of the CPT-test is the evaluation of the effect of the change in stress state on the strength of the soil by doing a test before and after, for example, ground improvement works.

2.2.2 Dilatometer test (DMT)

The flat dilatometer (DMT) is a stainless steel blade with a thickness of about 15 mm, having a flat, circular steel membrane (diameter 60 mm, thickness 0.1 to 0.2 mm) mounted on one side (**Figure 5**).

The blade is connected to a control unit on the ground surface by a pneumatic-electrical tube (transmitting gas pressure and electrical continuity) running through the insertion rods (same rods as for the CPT-test) (**Figure 6**). A gas tank, connected to the control unit by a pneumatic cable, supplies the gas pressure required to expand the membrane. At the beginning of the expansion, the horizontal pressure A can be measured. After an expansion of 1.1 mm at the center of the flexible membrane, the horizontal pressure B can be measured.

The pressure readings A and B must then be corrected by the values ΔA and ΔB, determined by calibration, in order to determine the soil stresses P0 (vertical position of the membrane) and P1 (expansion of 1.1 mm of the membrane). P0 is used for the determination of K_0, (P1-P0) is related to the stiffness of the soil.

The blade is pushed into the ground using common field equipment as for the CPT-test. The test starts by pushing the dilatometer into the ground. The blade is advanced into the ground of one depth increment (typically 20 cm), and the procedure for the A, B readings are repeated at each depth. A more detailed description of the test procedure is discussed in [4].

The field of application of the DMT is very wide. The immediate determination of the stress state and the stiffness of the soil are important advantages of this test technique.

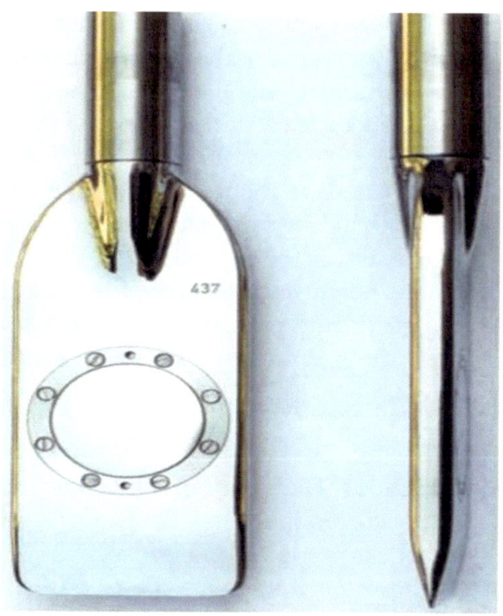

Figure 5.
Dilatometer blade.

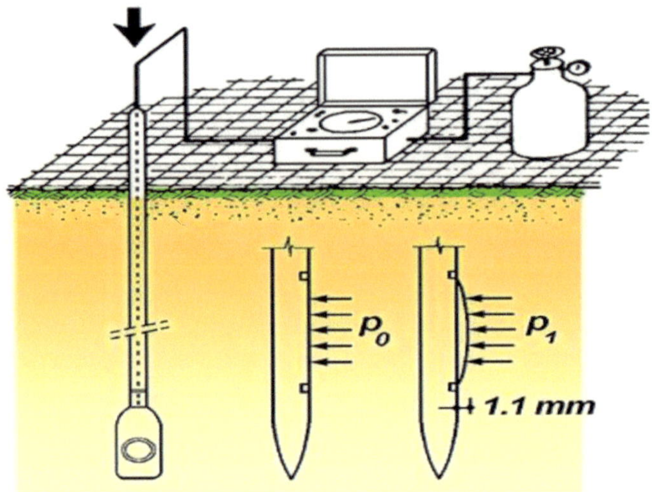

Figure 6.
DMT – Test procedure.

3. Instability of the studied slope

3.1 A general overview of the main sliding

In the next figures, a general view of the main sliding is presented. After excavation of the clay pit till a depth of about 30 m, in the Northern part of the pit, instability (small slidings) of the walls started to appears after some years. The main reason for this instability was the lack of sufficient resistance against horizontal stability. During the period of excavation, the pit became the central dewatering point for the region. Unfortunately, the surface between the upper quaternary layer and the underlying Boom clay was slightly declining towards the pit. The upper part (quaternary layer), with a thickness of about 4 to 5 m, moved towards the pit, resulting in instability. Also, the underlying Boom Clay was affected by this slide.

Figures 1 and 7 present a general view of the slides and the soil movement in 2014. As shown in these figures, a block of 100 m by 50 m in size started to move downward, resulting in a movement of about 9 m down at the corner of the pit.

In **Figure 2**, the extent of the sliding can be noticed. The legend in **Figure 2b** indicates the vertical displacement of the soil mass after the sliding.

After the first sliding in the northwestern part of the pit, a monitoring program was set up. The monitoring includes the execution of CPT-tests and DMT-test. Besides these tests, also the pore-water pressures were measured on a regular basis.

3.2 First sliding – period 2011 - 2012

The first slide occurred in 2011. The size of the slide was limited. In the second half of 2013, one more slide occurred in the neighborhood at point 5 shown in **Figure 2a.**, A photo of the slide is presented in **Figure 8**.

3.3 More pronounced instability of the slope (2014)

After a more pronounced sliding in 2013, in January 2014, a monitoring program was established for different points at the top of the slope around the pit. Besides topographic surveys and the measurement of the phreatic water level,

Figure 7.
Detailed impression of the instable slope (as adapted from Peiffer (2016) [5]).

Figure 8.
Photo of the unstable zone after the instability of the slope (as adapted from Peiffer (2016) [5]).

Cone Penetration Tests (CPT) and Dilatometer-tests (DMT) were carried out. The in situ tests were executed in points 1 to 5 (**Figure 2a**).

A lot of in situ tests were conducted. In this section, the results of the DMT tests at point 1, and point 2 (**Figure 2a**) were discussed, because the sliding occurred in this zone. For the general discussion of the results of the in situ tests can be referred to [5].

In June 2014, a big sliding occurred (size of about 100 m x 50 m) between points 1 and 2 (**Figure 2a**).

The extent of the affected zone can be seen in the **Figure 9**. This is a CAD-translation of the zone presented in **Figure 2a**. Two profiles (cross-sections) are also presented in **Figures 10** and **11**. The general Belgian reference system for the heights indicated as 'm TAW' refers to the height above sea level. The arrows on the figures refer to measurement points.

In July 2014, 1 month after the sliding, additional site investigation was done at the top of the slope in the area where the sliding occurred. Six additional points were defined (points A to F) for the execution of CPT and DMT tests.

4. Analysis based on the site investigation

As discussed in 2.2.2., The DMT test can be used for risk assessment concerning the stability of slopes [3]. In recent publications [5, 14], the results of CPT and DMT tests were interpreted in order to evaluate the sensitivity of these testing methods for the evaluation of slope stability.

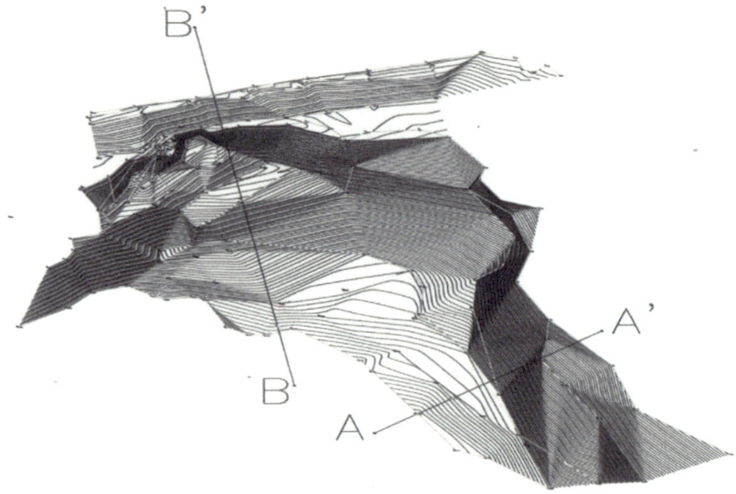

Figure 9.
Print of the NE-corner.

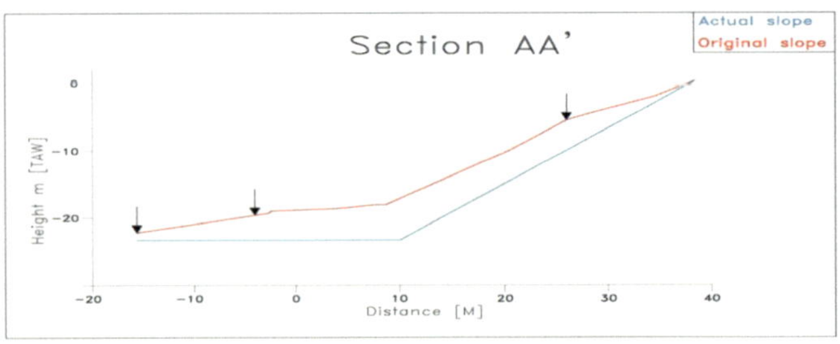

Figure 10.
Section A-A.

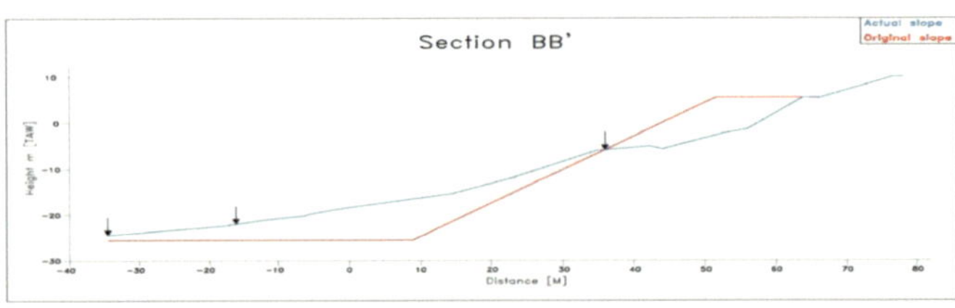

Figure 11.
Section B-B.

The consecutive DMT-measurements revealed an evolution towards instability. After a significant failure (sliding) in one zone in June 2014 (affecting an area of 100 m by 50 m), remedial works in the destabilized zone were undertaken (October – November 2014). After stabilization in 2015, further DMT-measurements were taken in order to investigate the evolution of the stress state of the soil in the potential sliding surfaces.

5. Boom clay

Based on geological investigations and data maps, the upper layer consists of sandy silt (quaternary deposit). Under this layer, the tertiary overconsolidated

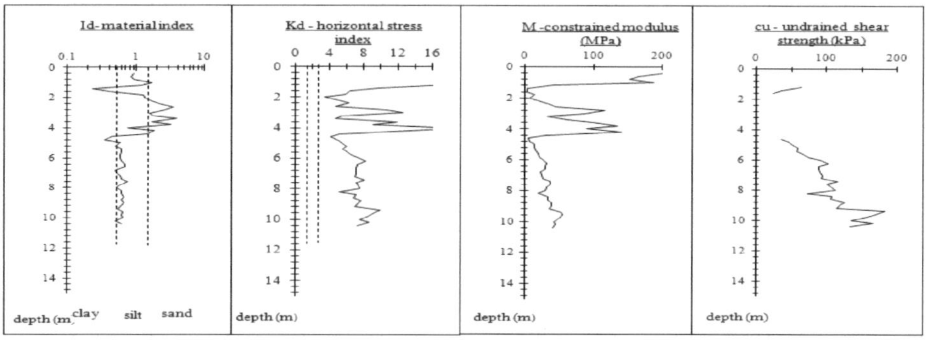

Figure 12.
DMT results in Kruibeke.

Boom Clay can be found. **Figure 9** shows the characteristics of this soil as measured by the DMT.

The tests allow for good identification of the soil. In the column on the left side, the results for the material index Id are presented. Based on Id, it is possible to make a reliable interpretation of the nature of the soil. One can identify in this graph a top layer of sand to silt and starting from a depth of about 4.5 m, a silty clay layer (the Boom Clay). In the second column, the horizontal stress index Kd is presented. A Kd-value between the two dashed lines (Kd = 1.8 and Kd = 2.4) refers to a normally consolidated soil. A Kd-value higher than 2.4 refers to an overconsolidated soil (OC), as can be seen for the Boom Clay.

Also, the stiffness (constrained modulus M) and the strength (undrained shear strength c_u) can be determined out of the basic measurements of the DMT. Both an undrained shear strength of 100 to 150 kPa and a constrained modulus up to 35 MPa refer to very stiff clay.

The profile confirms the results of the undisturbed sampling of the soil. This section focused on the horizontal stress index Kd. This parameter reflects the ratio of horizontal stress to vertical stress (**Figure 12**).

The phreatic level is at a depth of about 1.5 m. The thickness of the clay at the bottom of the pit is about 10 m. This thickness is sufficient to resist the upward artesian water pressures.

During the period of excavation, the pit became the central drainage point for the region. Unfortunately, the surface between the upper quaternary layer and the underlying Boom clay was declined slightly towards the pit.

6. DMT as a tool for the detection of slip surfaces – the DMT-Kd-method

The principle of the DMT-Kd-method for the detection of slip surfaces have been discussed by many authors [3, 13].

In many OC clay landslides, the sequence of sliding, remolding, and reconsolidation leave the clay in the slip zone in a normally consolidated (NC) or nearly NC state, with one or more loss of structure, aging or cementation.

Based on field data from different clay sites in various geographical areas, correlations between OCR and Kd could be established, as shown in **Figure 13** [2, 13, 15–18].

In genuinely NC clay (no structure, aging or cementation), the horizontal stress index Kd from the DMT is approximately equal to 2, while Kd values in OC clays are considerably higher (for the Boom Clay about 8).

For this reason, it is known that, if an OC clay slope contains clay layers with Kd ≈ 2, these layers are highly likely to be part of a slip surface (active or quiescent).

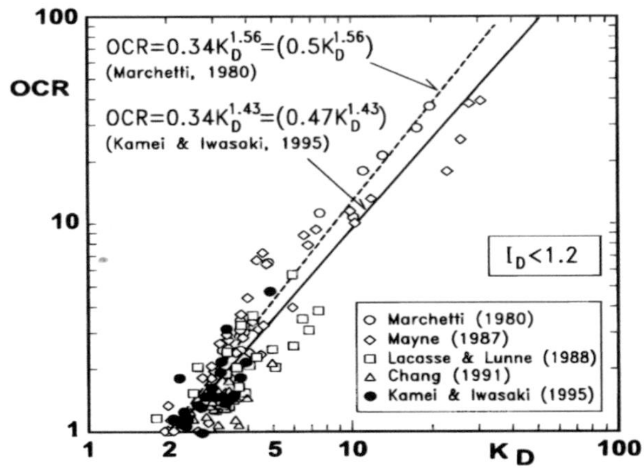

Figure 13.
Correlation Kd-OCR for cohesive soils [2, 13, 15–18] (as adapted from Peiffer (2016) [5]).

The DMT-Kd-method consists of identifying zones of NC-clay in a slope, using Kd ≈ 2 as the identifier of the NC zones.

7. Evaluation of the test results

The detailed evaluation of the test results of profiles with depth is discussed in [14, 19].

In **Figure 14**, the results are presented for the tests before (January 2014 – DMT 1), immediately after (July 2014 – DMT E) and about 2 years after the sliding (March 2016 – DMT 1′). In the test diagrams, the presence of the upper quaternary sandy silt layer can clearly be distinguished. Between January and July 2014, Kd decreased from about 3 to 5 t0, about 1.2 to 1.8 between 3.5 m and 4.5 m of depth. One can conclude that at this depth, the soil modified from an overconsolidated soil to an unstable soil (horizontal stress state lower than that for a normally consolidated soil). Above and below this strongly altered zone, the Kd-value decreased also, but less pronounced. Also, the stiffness, in particular the constrained modulus M, reduced sharply. The CPT diagram also shows a reduction of the cone resistance at a depth of about 3 to 4 m, but not that pronounced as in the results of the DMT, especially not at the top of the overconsolidated clay.

Out of the results of the DMT-tests, one can observe an unstable layer with a thickness of about 1.5 m where the soil is unstable where the slip was developed, but above and below this zone, the stress state is also affected by the sliding. Although cone resistance is decreased by a factor about 2, and a significant decrease of qc between 2 and 4 m, it can be seen in this layer that the qc-profile does not give a clear reflection of the sliding. It is also important to note that the Kd-values below the sliding zone decreased to a depth of about 5 m below the sliding surface.

It is worth noting how and to what extent the stresses were changed after the sliding. In **Figure 14c**, the results are presented for the measurements carried out in March 2016, after remedial works had been carried out in October–November 2014. The qc-value did not change significantly. The Kd-diagram shows clearly that the Kd-values exceeded again 1.8 (the lower limit for normally consolidated soils). For the clayey layer, the reading actually reflects a normally consolidated soil. For the intermediate sandy silt layer, one can see a slightly overconsolidated soil condition.

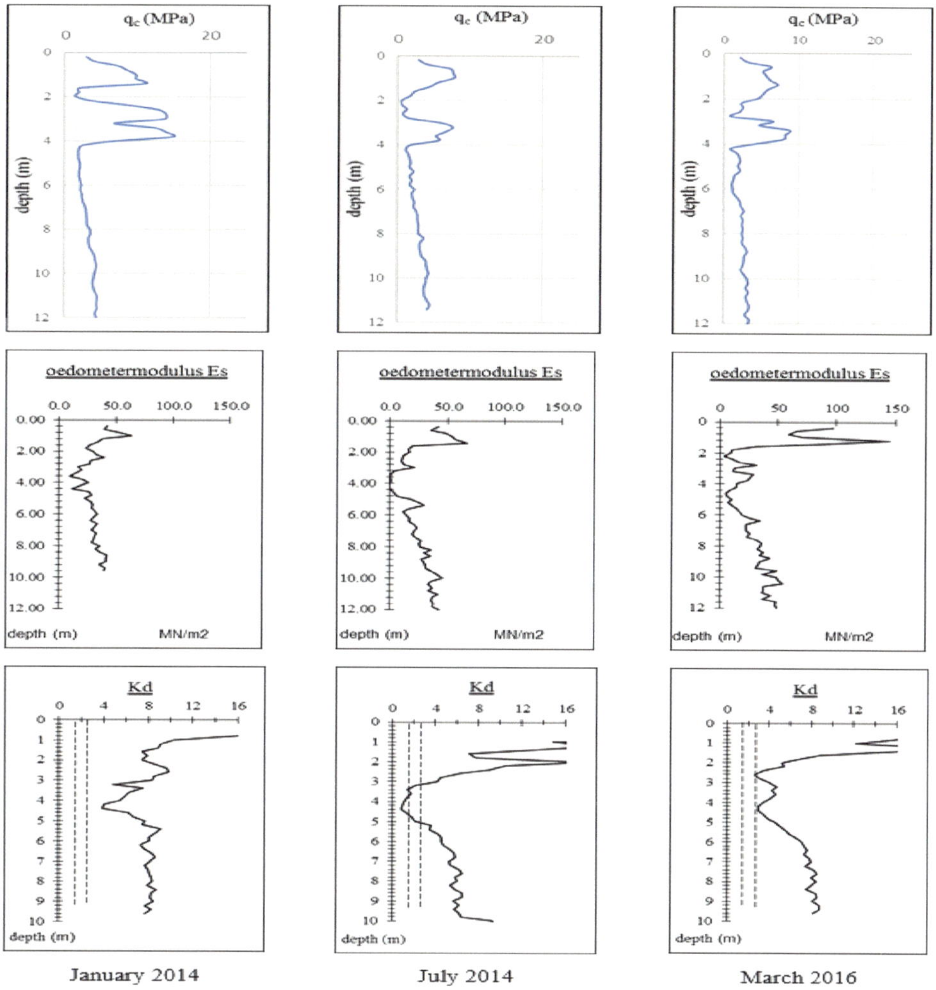

Figure 14.
Comparison CPT-DMT results – (a)January 2014 (DMT 1); (b) July 2014 (DMT E) and (c)march 2016 (DMT 1') (as adapted from Peiffer (2016) [5]).

8. Stability analysis

Based on the field measurements, the origin and the propagation of the instable zone where the sliding surface was initiated and developed could be clearly investigated.

A numerical analysis was performed, considering the geometry of section AA' (**Figure 10**) and the soil characteristics as they were measured in situ (before the sliding) and in laboratory tests. In this analysis, variable piezometric boundary conditions were considered to evaluate the impact on stability. Further investigations and additional calculations showed the importance of the groundwater pressures on the horizontal equilibrium.

Additional CPT-tests and laboratory tests were executed in order to make a detailed stability analysis. Based on an analytic and numerical analysis (PLAXIS-calculation), it was possible to prove the cause of the instability, and to design an appropriate improvement of horizontal stability.

Apparently, during the period of exploitation, the environment around the pit gradually became a central dewatering point. The flow of water towards the pit increased, and also the flow gradient increased.

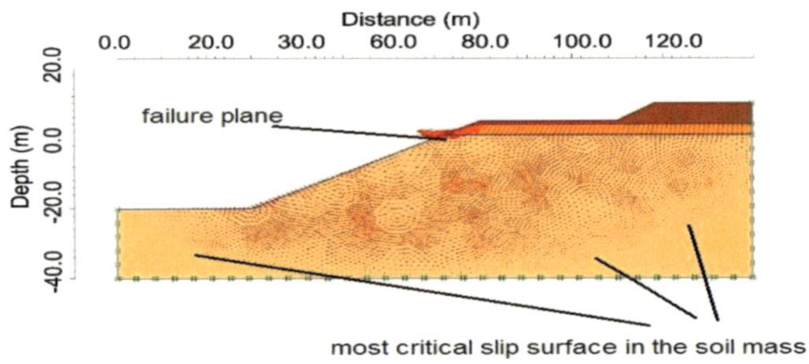

Figure 15.
PLAXIS-analysis and measurement of the slope after sliding (as adapted from Peiffer (2016) [5]).

Figure 15 [19] presents the calculated critical horizontal disequilibrium taking into account the presence of the groundwater and the piezometric site conditions. The red dots are points where the soil becomes plastic. The higher the density of the red dots, the more plastic the soil will behave, finally resulting in failure when there is no equilibrium of the soil stresses anymore. At the failure (the dark red line in **Figure 15**) zone, one can clearly detect the position of the sliding plane.

Out of these theoretical results, it became clear that the installation of drains would play an important role in rehabilitation.

9. Remediation

The slope was stabilized using three techniques:

- by installing deep drains in order to reduce the water stresses
- by smoothening the slope
- by the construction of a piled barrier

The PLAXIS-analysis resulted in a design where a deep drain has to be installed at a depth of 5 m., as shown in **Figures 16** and **17**. This drain was installed at the beginning of November 2015. The red line in **Figure 15** is the position of the drain in the plan view.

After the installation of this drain, there was an immediate effect on the depth of the groundwater. The original depth of 1.5 m increased to a depth of about 4.5 m. The safety factor of the slope stability increased from 0.9 to 1.31.

The original angle of the slope (before sliding) was 27° After the sliding, the clay was remolded.

Before the rehabilitation works of the slope, an angle of the slope of 18° was realized. These works were realized in order to reduce the risk of additional slidings. The result of this work is presented in **Figure 18.**

Although the slope angle was reduced, the soil mass remained moving to the center of the pit. It was not possible to remove the soil because of the bad soil conditions.

In order to stop the movement of the sliding mass, a piled barrier (6 rows of piles) was installed (**Figure 19**).

Between the piled barrier and the original slope, vertical drains were installed for the evacuation of water during consolidation.

Figure 16.
Installation of the deep drainage.

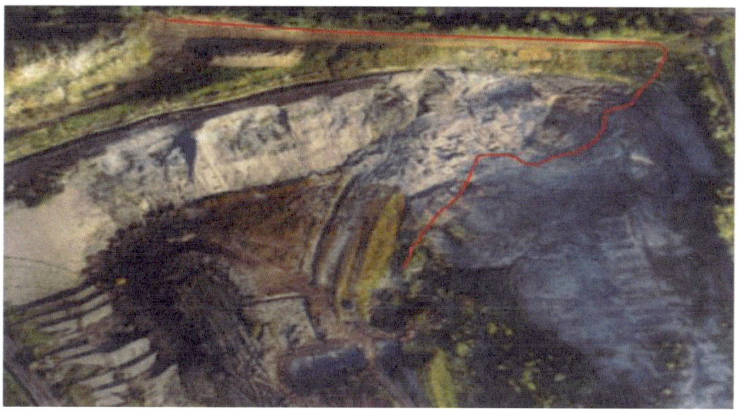

Figure 17.
Position of the drain in plan view (red line).

Figure 18.
Smoothening of the slope.

Since the beginning of 2019, the rehabilitation of the slope has been started. These works are executed in successive phases in order to guarantee the stability of the site during the works. Each phase has a height of 3 to 4 m, as can been seen on **Figure 20**.

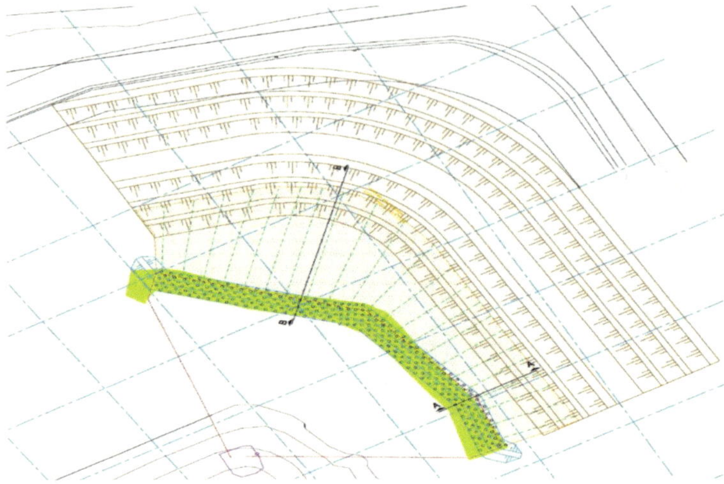

Figure 19.
Piled barrier.

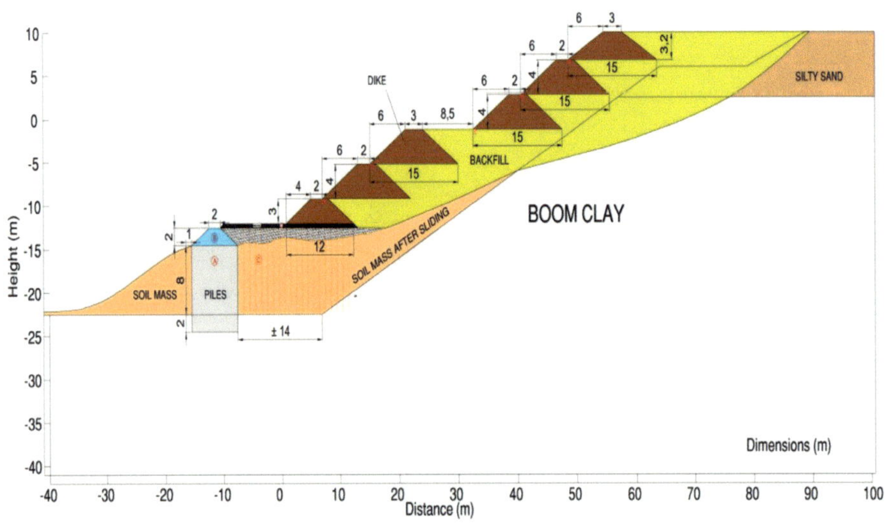

Figure 20.
New profile of the slope.

For each phase first a dike is constructed with well bearing sand. Between the dike and the actual slope, a backfill is made with sandy soil.

Today, about 50% of the total work has been completed.

10. Conclusions

In this chapter, DMT is applied as a tool for the monitoring of the stress state in a slope before and after the sliding.

From an analysis of the sliding observed in the overconsolidated Boom Clay, it can be concluded that it is possible to measure the effect of sliding on soil conditions and to evaluate the stability of the soil. In addition, it is possible to detect the position of a slip surface and the zone, which is influenced by the sliding.

For both applications, the results of a CPT-test in this type of soil are, as could be expected, less sensitive to changes in a stress state.

Acknowledgements

This research was possible with the funding by Alpha-Engineering.

Author details

Herman Peiffer
Ghent University, Zwijnaarde, Belgium

*Address all correspondence to: herman.peiffer@ugent.be

References

[1] Bozzanoa, F., Bretschneider, A., Martinoa, S., Prestininzia,A. Time variations of the K0 coefficient in overconsolidated clays due to morphological evolution of slopes. Engineering Geology,Vol. 169; 2014; p.69-79.

[2] Chang, M.F. Interpretation of overconsolidation ratio from in situ tests in recent clay deposits in Singapore and Malaysia. Canadian Geotechnical Journal, Vol. 28 n°; 1991. p. 210-225

[3] Totani, G., Calabrese, M., Marchetti, S., Monaco, P. Use of the in-situ flat dilatometer (DMT) for ground characterization in the stability of slopes. XV ICSMFE; Hamburg; Vol. 1; 1997. p. 607-610.

[4] Marchetti, S., Monaco, P., Totani, G., Calabrese, M. The Flat Dilatometer Test (DMT) in soil investigations. A Report by the ISSMGE Committee TC16, International Conference On In Situ Measurement of Soil Properties; Bali (Indonesia); 2001.

[5] Peiffer, H. A comparative analysis of the stability of a slope in an overconsolidated clay pit based on CPT and DMT measurements, Fifth International Conference on Geotechnical and Geophysical Site Characterisation, Australian Geomechanics Society, Sydney, Vol.2; Queensland, Australia; 2016; p. 1359-1364.

[6] Doran, I.G., Sivakumar, V., Graham, J, Johnson, A. Estimation of in situ stresses using anisotropic measurements and suction measurements. Geotechnique, 50 (2); 2000; p.189-196.

[7] Sivakumar, V., Doran, I.G., Graham, J., Navaneethan, T.Relationship between K0and overconsolidation ratio: a theoretical approach. Geotechnique, 52 (3); 2001; p.225-230.

[8] Sivakumar, V., Navaneethan, T., Hughes, D., Gallagher, G. An assessment of the earth pressure coefficient in overconsolidated clays. Geotechnique, 59 (10); 2009; p. 825-838.

[9] Silvestri, V. Long-term stability of a cutting slope in an overconsolidated sensitive clay. Canadian Geotechnical Journal, Vol.17, N3; 1980; p. 337-351.

[10] Grefsheim, F.D. Laboratory testing for slope stability design parameters in overconsolidated clay. Proc 5th International Symposium on Landslides, Lausanne (Publ. A.A. Balkema); 1988; p. 169-174.

[11] Chandler, R.J. The long-term stability of cutting slopes. Geotechnique, 24; 1974; p.21-38.

[12] Dehandschutter, B. ; Vandycke, S., Sintubin, M.,Vandenberghe, N. Wouters, L. Brittle fractures and ductile shear bands in argillaceous sediments: inferences from Oligocene Boom Clay (Belgium). Journal of Structural Geology, Vol. 27; 2005; p. 1095-1112.

[13] D'Elia, B. Esu,F. Marchetti,S. and Totani,G. In situ DMT testing and observations in landslide areas in overconsolidated clays. In:Proc. 7th ISL; 17-21 June 1996; Trondheim, Norway. p. 1607-1612.

[14] Peiffer, H. The use of a DMT to monitor the stability of a clay exploitation pit in the Boom Clay in Belgium. 3rd International Conference on the flat Dilatometer; Rome, Italy; 2015; p. 127-133.

[15] Marchetti, S. In Situ Tests by Flat Dilatometer. J. Geotech.Eng. Div., Vol. 106, No. 3; 1980; pp. 299-321.

[16] Kamei, T., Iwasaki, K. Evaluation of undrained shear strength of cohesive soils using a flat dilatometer. Soils and Foundations 35 (2); 1995; p. 111-116.

[17] Lacasse, S. and Lunne, T. Calibration of dilatometer correlations", Proc. Isopt-1 1; 1988; p. 348-359.

[18] Mayne, P. Determining Preconsolidation Stress and Penetration Pore Pressure from DMT Contact Pressures. Geotech. Test. J., Vol. 10, No. 3; 1987; pp. 146-150.

[19] Peiffer, H. Studierapport stabiliteit kleigroeve te Kruibeke. Project Report; Antwerpen; Belgium; 2015.

Chapter 9

Landslide Potential Evaluation Using Fragility Curve Model

Yi-Min Huang, Tsu-Chiang Lei, Bing-Jean Lee and Meng-Hsun Hsieh

Abstract

The geological environment of Taiwan mainly contains steep topography and geologically fragile ground surface. Therefore, the vulnerable environmental conditions are prone to landslides during torrential rainfalls and typhoons. The rainfall-induced shallow landslide has become more common in Taiwan due to the extreme weathers in recent years. To evaluate the potential of landslide and its impacts, an evaluation method using the historical rainfall data (the hazard factor) and the temporal characteristics of landslide fragility curve (LFC, the vulnerability factor) was developed and described in this chapter. The LFC model was based on the geomorphological and vegetation factors using landslides at the Chen-Yu-Lan watershed in Taiwan, during events of Typhoon Sinlaku (September 2009) and Typhoon Morakot (August 2009). The critical hazard potential (*Hc*) and critical fragility potential (*Fc*) were introduced to express the probability of exceeding a damage state of landslides under certain conditions of rainfall intensity and accumulated rainfall. Case studies at Shenmu village in Taiwan were applied to illustrate the proposed method of landslide potential assessment and the landslide warning in practice. Finally, the proposed risk assessment for landslides can be implemented in the disaster response system and be extended to take debris flows into consideration altogether.

Keywords: landslide, fragility, landslide potential, probabilistic model

1. Introduction

Taiwan is on the path of western Pacific typhoon path and on the circum-Pacific earthquake belt, indicating that Taiwan suffered from two or more natural disasters, which was the highest in the world [1]. Besides, most of the land in Taiwan, about 70% of total area, is hillside. Given the conditions of increasing impacts of climate change and extreme weathers, the rainfall-induced landslide has become a serious issue in Taiwan.

Most landslide researches used the landslide susceptibility analysis (LSA) to develop landslide evaluation model [2]. The LSA models basically use factors and observed data to construct the description of landslides. The factors include rainfall intensity, accumulated rainfall, slope degree, vegetation, etc. The common models developed for landslide hazard or landslide evaluation are usually deterministic analysis, including the traditional slope stability analysis [2]. Recently, a novel concept of applying probability to landslide evaluation had been proposed.

The fragility curves, which are commonly used in the earthquake-induced structure analysis, had been adopted to represent the probability of landslide [3–5]. The process of applying fragility curve to landslide evaluation is to consider and estimate the recurrence and the probability of exceedance of a damage level for a landslide [3, 4].

In this chapter, the preparation of landslide fragility curves was introduced. The procedure of developing the landslide fragility curve (LFC) model was the researches of rainfall-induced shallow landslide in the past years [2–5]. The proposed LFC model considered the impacts of rainfall and the vulnerability of environment. Instead of using one-variable triggering factor (rainfall intensity or accumulation) in the previous research [2], the newly improved LFC model used bivariate approach in the model [3, 4]. The improved LFC model introduced the landslide fragility surface (LFS) by considering the influence of both rainfall intensity and accumulation at the same time [4, 5].

The spatial statistics and geographic information system (GIS) were used for data processing. The data of each factor used in the model was further divided into groups. Classification of factors represented the environmental characteristics of a specific area. The analysis basis was conducted spatially on the slope units, which are topographically defined as the parts of a watershed [5]. With the LFS model, the risk assessment of landslide then was analyzed in association with the rainfall hazard potential [4, 5]. The Shenmu area of Chen-Yu-Lan watershed was selected as the study area, and historical cases were used to illustrate the application of LFS model.

2. The factors and environmental database

When considering the factors to be used in the landslide problem, these factors are generally classified as triggering and environmental factors [6–8]. Among these factors, the rainfall is usually the major concern, and for environmental vulnerability, many factors can be chosen from. Not every chosen environmental factor can be used in developing a landslide model because of (1) few data in the database, (2) lack of data, and (3) low influence in the model. In this chapter, the cumulative rainfall and maximum hourly rainfall (rainfall intensity) were used for triggering factors, whereas slopes, slope aspects, landslide area, incremental landslide area, ratio of incremental landslide area, normalized difference vegetation index, distance to the nearest river, and geology were used for environmental factors of hillside slope in the study. A GIS database to describe landslide areas was created and was later applied in developing the proposed fragility curve model. These indexes, factors, and symbol definitions are explained in the following:

1. Maximum rainfall intensity (I_{max}): the maximum rainfall intensity is the rainfall in the form of rainfall per unit time. In this study, I_{max} refers to the maximum hourly rainfall (**Figure 1**) and was used as a triggering factor for LFC model.

2. Effective accumulated rainfall (R_{te}): the R_{te} is defined as the accumulated rainfall before the maximum rainfall intensity in a continuous raining event (**Figure 1**), by considering the influence of antecedent 7-day rainfall.

3. Hillside slope (S): the dynamic behavior of the landslide has close relationship with the slope. Hence, the degree of slope may be a prominent factor of triggering landslides. In this study, the slope was classified based on the Soil and

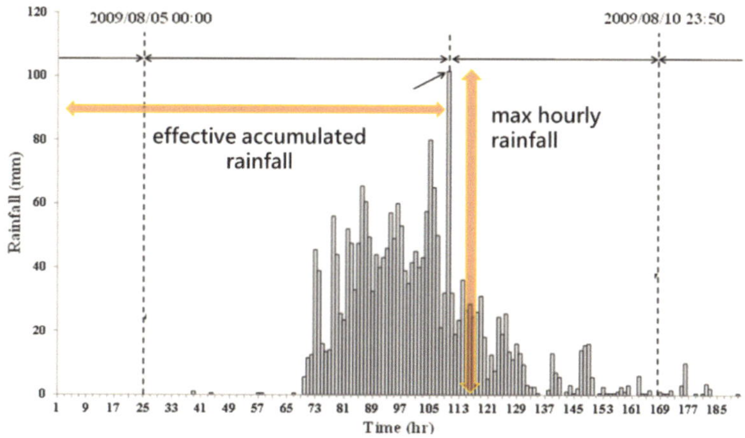

Figure 1.
The definition of rainfall indices: I_{max} and R_{te} (modified after [2–4]).

Water Conservation Bureau manual [9]. There are seven slope levels of 5% or less, 5–15%, 15–30%, 30–40%, 40–55%, 55–100%, and slope exceeding 100%. The slopes <15% are recognized as flat ground or very gentle slopes and not included in this study. Slopes of levels 3–7 were studied in the landslide model.

4. Slope aspect (A): the slope aspect represents the vulnerable directions of occurring landslide when given a known topography. This factor may represent the "weak" aspect of a slope in terms of landslide.

5. Landslide area (LA): observing the landslide distribution through image classification results can obtain the information about the land cover change. The change from events of Typhoon Sinlaku (in 2008) and Typhoon Morakot (in 2009) was identified using GIS software.

6. Incremental landslide area (IA): to understand the landslide increment, the images before and after a landslide were considered. The landslides are classified into five categories (shown in **Figure 2**): (1) the original landslide area (number 1 + 2), (2) the original landslide area extension (number 2), (3) new landslide area on single period (number 3), (4) new landslide area on pre-/post periods (number of 2 + 3), and (5) vegetation restoration area (number of 1). In this study, the new landslide area on pre-/post periods (number of 2 + 3) was considered.

7. Ratio of incremental landslide area (RIL): to obtain the ratio of incremental landslide area, this study used the incremental landslide area from image of two periods to determine this factor.

8. Vegetation index (N): to determine the density of vegetation on a patch of land, researchers must observe the distinct colors (wavelengths) of visible and near-infrared sunlight reflected by the plants [10]. Nearly almost satellite vegetation indices employ the difference formula, $(NIR - R)/(NIR + R)$ [11], to quantify the density of plant growth on the earth—the subtraction of near-infrared radiation (NIR) and red radiation (R) divided by the addition of near-infrared radiation and red radiation. The result of this formula is called the normalized difference vegetation index (NDVI). The values for NDVI in this study were obtained from SPOT image. The range of NDVI is −1 to 1.

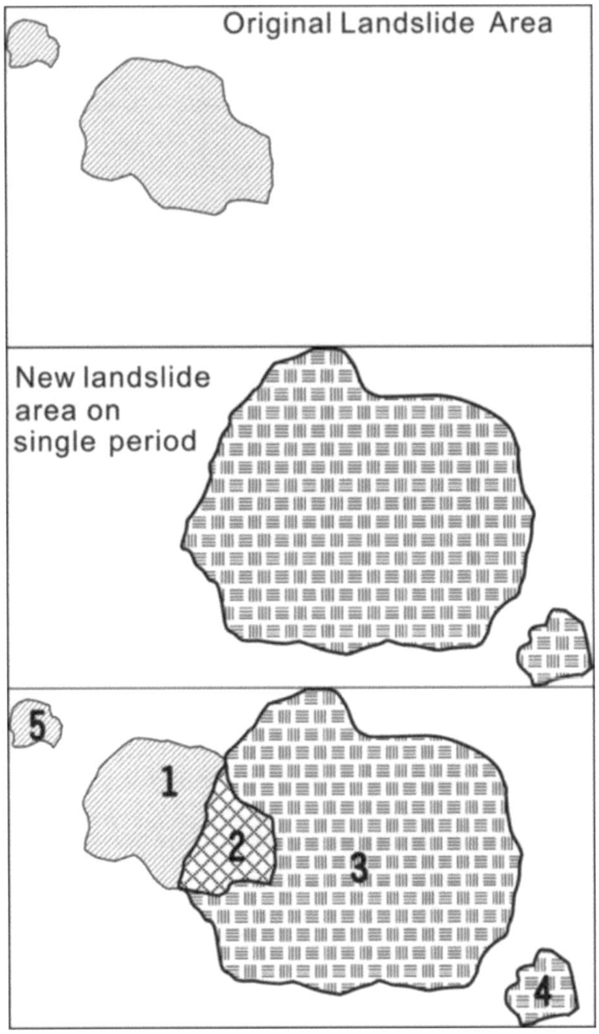

Figure 2.
Concept of mapping landslide area change: differences between two periods of SPOT image [2].

9. Distance to the nearest river (R): the landslide may be triggered due to the erosion by the river at the toe section. The distance to the river reflects the potential of landslide contributed from the river system.

10. Geology (G): the geological time scale of the area and the rock types of the site were combined into consideration as the geology factor. In the past studies, the geology-related information (like the rock types and rock strength) was not usually available. Therefore, to simplify the classification, the geological time scale was chosen to represent the possible influence of geology.

3. Study area and material

To explain the landslide fragility model, the Shenmu area in Taiwan was used as a case to demonstrate the development of LFC of a given site. The Shenmu area locates in the watershed of Chen-Yu-Lan River. Chen-Yu-Lan watershed is at the central part of Taiwan (**Figure 3**). The Chen-Yu-Lan River originates from the

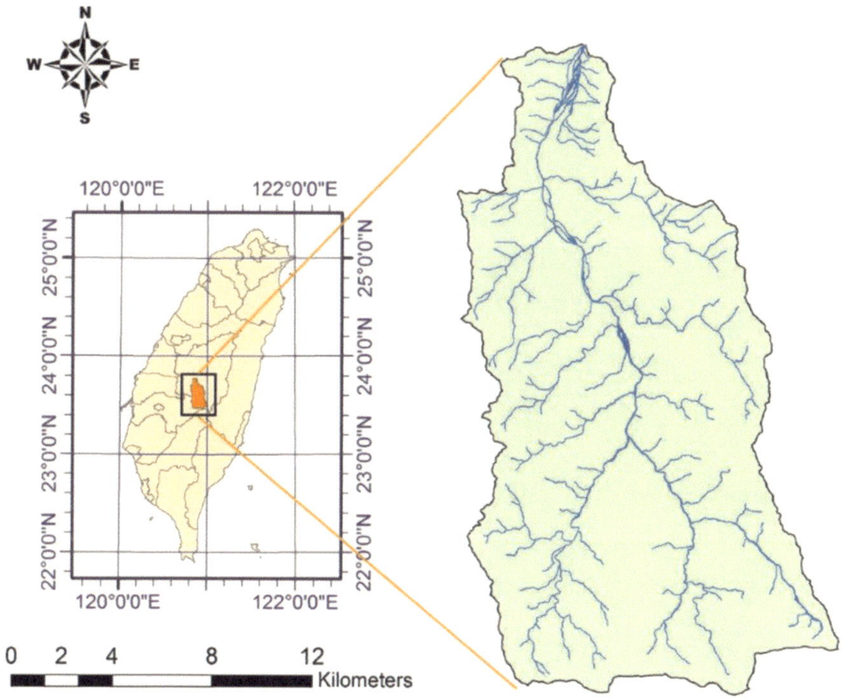

Figure 3.
Chen-Yu-Lan watershed [2].

north peak of Yu Mountain and is one of the upper rivers of the Zhuoshui River system, which is the largest river system in Taiwan. Chen-Yu-Lan River has a length of 42.4 km with an average declination slope of 5%, and its watershed area is about 450 km^2. This area was fragile after the Chi-Chi Earthquake (occurred on September 21, 1999).

The Shenmu area is a location where debris flows frequently occurred [5]. The local village is adjacent to the confluence of three streams: Aiyuzi Stream (DF226), Huosa Stream (DF227), and Chushuei Stream (DF199). In Shenmu, the debris flows usually occurred at the Aiyuzi Stream due to its shorter length and large landslide area (**Table 1**) in its upstream [5]. **Figure 4** shows the terrain of three streams.

In addition to the basic terrain data of Shenmu area, the hydrologic and geographic factors are needed in modeling. To obtain these factors, an environment database of Chen-Yu-Lan watershed was prepared. Among the data collection, the landslide increment (i.e., new landslides) after a rainfall event was also obtained by image processing method in this study.

To develop the LFC model, the local environmental data was collected for the study area, and GIS was used to process the data. The environment database of Chen-Yu-Lan watershed includes data of geology, geological layers, rock property, slope and slope aspects, and DEM, as shown in **Figures 5–8**.

Debris flow no.	Stream	Length (km)	Catchment area (km^2)	Landslide area (km^2)
DF199	Chushuei stream	7.16	8.62	0.33
DF227	Huosa stream	17.66	26.20	1.49
DF226	Aiyuzi stream	3.30	4.00	1.00

Table 1.
The landslide area in Shenmu after 2009 [5].

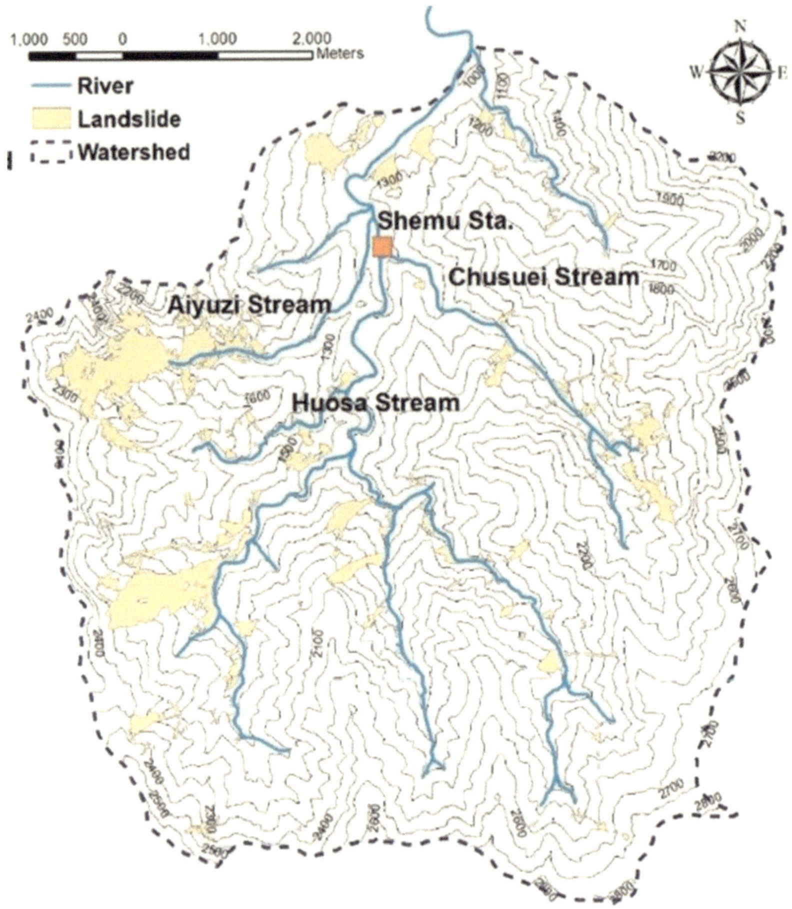

Figure 4.
The terrain and landslide areas of Shenmu area.

The new landslide areas (**Figures 9** and **10**) were identified by using pre- and post-event satellite images of Typhoon Sinlaku in 2008 and Typhoon Morakot in 2009 (**Table 2**). These landslide areas were used for later LFC model analysis. Another important factor in the LFC model is the vegetation conditions. The information of vegetation status was also obtained by image processing the same as the determination of new landslides.

In addition to the hydrologic and geographic data, the landslide triggering factors were also considered in data preparation. **Table 3** defines the rainfall indices. It should be noted that the effective accumulated rainfall was calculated by including the antecedent 7-day accumulated rainfall. The antecedent 7-day accumulated rainfall is the total weighted rainfall counted from the 7-day duration before the starting of current rainfall event. Take Typhoon Sinlaku (September 11–16, 2008) for example. The starting date of Typhoon Sinlaku was September 11, 2008, and the antecedent 7-day accumulation rainfall was the total weighted rainfall during September 3 to September 10, as described as *Ra* in **Table 3**.

Figures 11 and **12** show the rainfall interpolation of the events of Typhoon Sinlaku (September 11–16, 2008) and Typhoon Morakot (August 5–10, 2009). The red spots in the figure are the locations of rainfall stations. It was noted that the rainfall intensity and the cumulative rainfall of event of Typhoon Morakot were much higher than those of Typhoon Sinlaku. Both events had caused serious landslides in the central Taiwan.

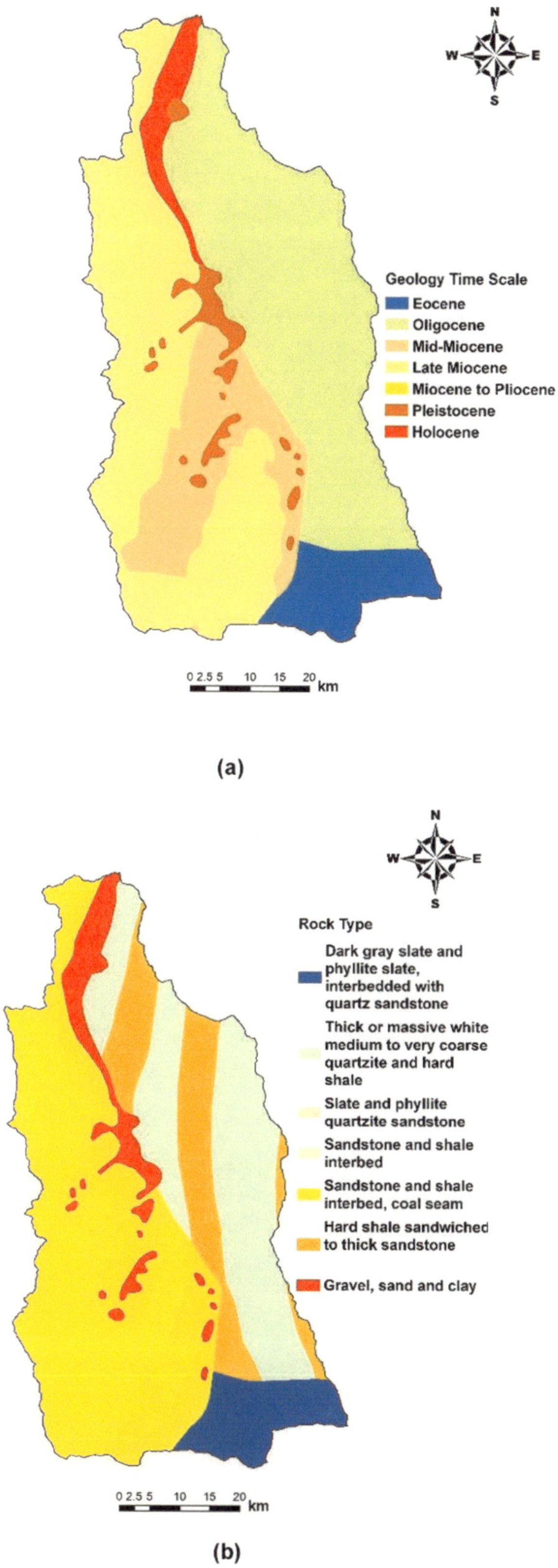

Figure 5.
Chen-Yu-Lan watershed: (a) geological time scale and (b) rock types.

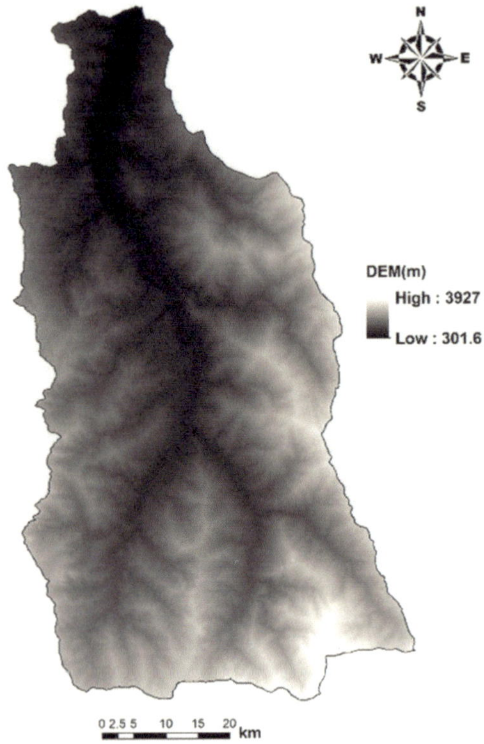

Figure 6.
Five-meter DEM of Chen-Yu-Lan watershed (after [2]).

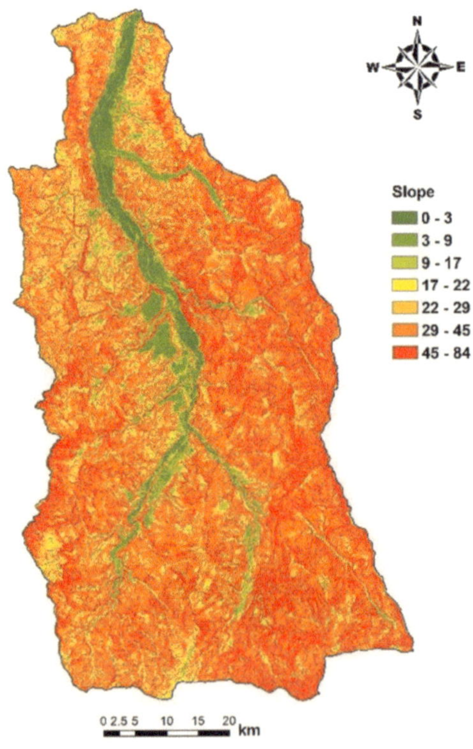

Figure 7.
The slope of Chen-Yu-Lan watershed.

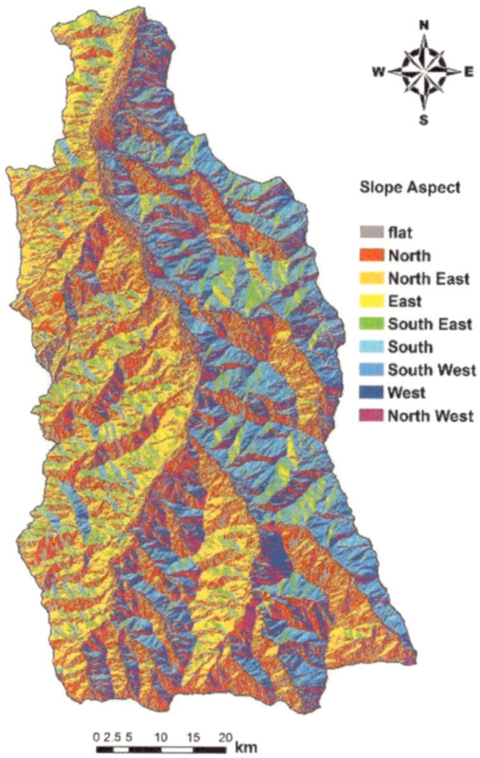

Figure 8.
The slope aspects of Chen-Yu-Lan watershed.

Finally, the database was used to analyze the study area on the basis of slope units. The slope unit was defined as in **Figure 13**. A slope unit is defined as one slope part or the left/right part of a watershed. Slope units can be topologically divided by the watershed divide and drainage line, with the help of GIS tool [12]. The application of slope unit in the development of LFC was based on the physical interpretation of slopes in the mountain area. The environmental database was applied in accordance with the slope units at the site of interest. **Figure 14** shows the slope unit distribution (total 5872 units) of Chen-Yu-Lan watershed.

4. Development of empirical landslide fragility model

To develop the empirical landslide fragility model, a probability distribution was chosen to describe the potential of landslide fragility. When the probability distribution was determined, the parameters of probability, the median and standard deviation, were obtained by fitting the data from the environmental database and the landslide areas. The use of slope unit was adopted here, and the classification of environmental factors was applied to represent the conditions of landslide given rainfall intensity and accumulated rainfall. The procedure of developing the empirical landslide fragility curve was described in the following.

4.1 Probability distribution of LFC

The fragility analysis is usually used to describe the potential of hazard in terms of potential levels or probability of exceedance of a level. To describe the probability

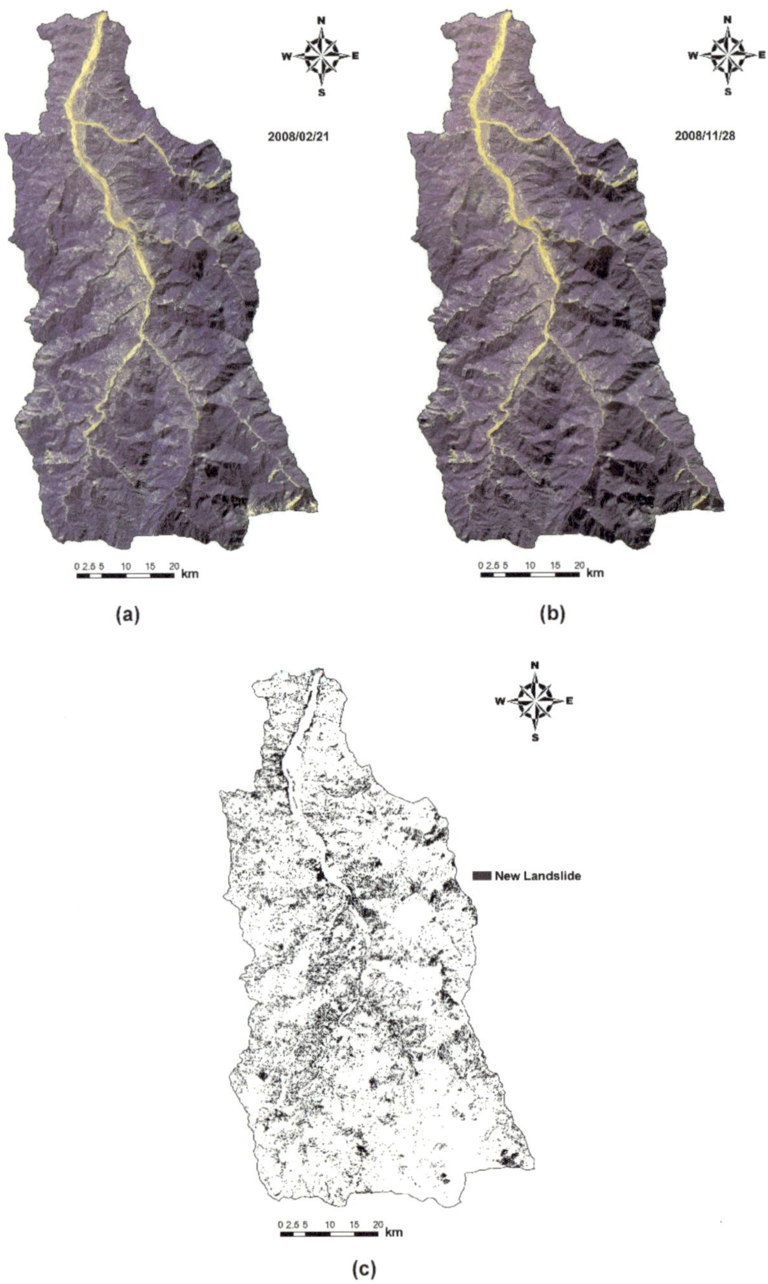

Figure 9.
Satellite images of pre- (a) and post-event (b) Typhoon Sinlaku and the new landslide areas (c) in Chen-Yu-Lan watershed.

about a hazard fragility, a feasible probability distribution can be assumed and applied in the model. The fragility curve of landslide, therefore, was assumed to be a lognormal distribution [12, 13]. The lognormal distribution can be constructed simply by the values of median and lognormal standard deviation and are called bivariate parameters (Eq. (1)):

$$f_j\left(x; c_j, \zeta_j\right) = \frac{1}{\sqrt{2\pi}\zeta_j x} e^{-\frac{1}{2}\left(\frac{ln\left(x/c_j\right)}{\zeta_j}\right)^2} \tag{1}$$

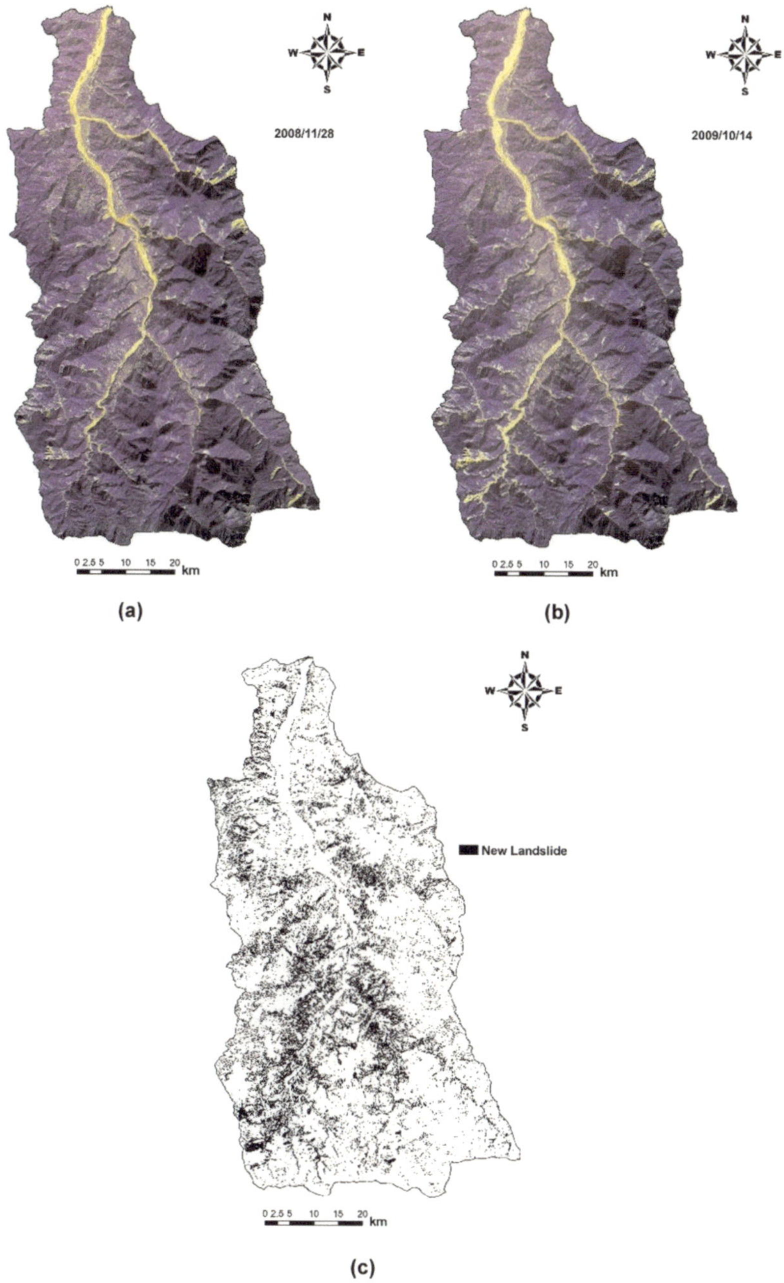

Figure 10.
Satellite images of pre- (a) and post-event (b) Typhoon Morakot and the new landslide areas (c) in Chen-Yu-Lan watershed.

where f_j is the probability density function of lognormal distribution, c_j is the median, ζ_j is the log-standard deviation, and x is the variable. The cumulative distribution of Eq. (1) is used as the fragility curve. The cumulative density function of lognormal distribution is expressed as Eq. (2):

$$F_j\left(x;c_j,\zeta_j\right) = \frac{1}{2} + \frac{1}{2}\,erf\left[\frac{ln\left(x/c_j\right)}{\zeta_j\sqrt{2}}\right] \tag{2}$$

Watershed	Event	Image time	Satellite	Incremental area (km^2)
Chen-Yu-Lan 448.14 km^2	Pre-Sinlaku	February 21, 2008	SPOT5	9.52 (2.12%)
	Post-Sinlaku	November 28, 2008	SPOT5	
	Pre-Morakot	November 28, 2008	SPOT5	10.21 (2.28%)
	Post-Morakot	October 14, 2009	SPOT5	

Table 2.
Satellite images of events at Chen-Yu-Lan watershed.

Index	Symbol	Definition
Max. hourly rainfall	I_{max}	The maximum hourly rainfall in a rainfall event
Effective accumulated rainfall	R_{te}	The antecedent 7-day accumulated rainfall (with reduction factor of 0.7*) before the starting of current event and the accumulated rainfall before the max. hourly rainfall in current event

**Antecedent 7-day accumulated rainfall (R_a) can be calculated by $R_a = \sum_{i=1}^{7} 0.7^i R_i$, where R_i is the daily rainfall of the ith day before.*

Table 3.
The rainfall indices.

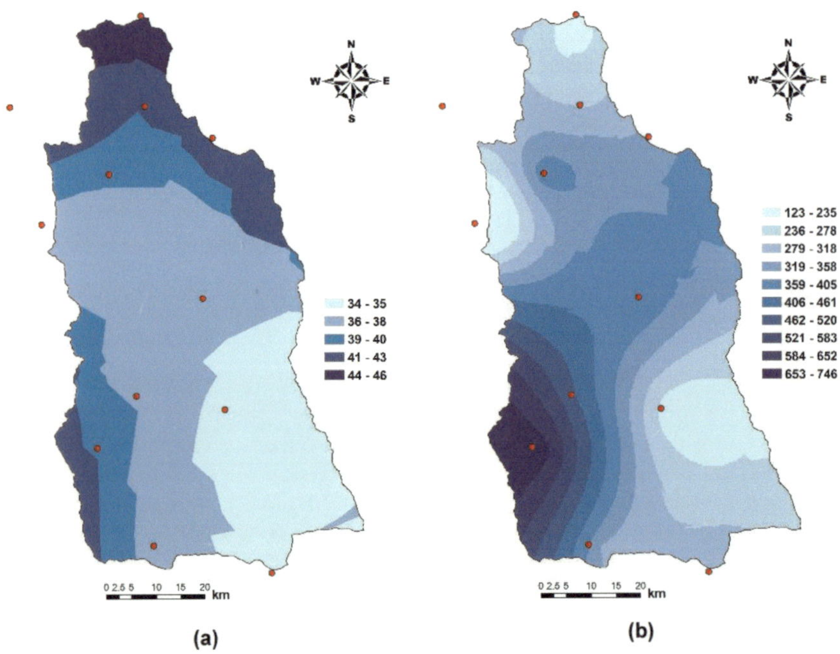

Figure 11.
Rainfall indices of Typhoon Sinlaku: (a) I_{max} and (b) R_{te}.

Eq. (2) represents the *j*th fragility, and *erf()* is the Gaussian error function. When the median and log-standard deviation are determined, the fragility curve of *j*th level can be obtained. The maximum likelihood estimation (MLE) can be applied to determine the median and log-standard deviation [13]. The aforementioned equations are suitable for one-variable estimation model.

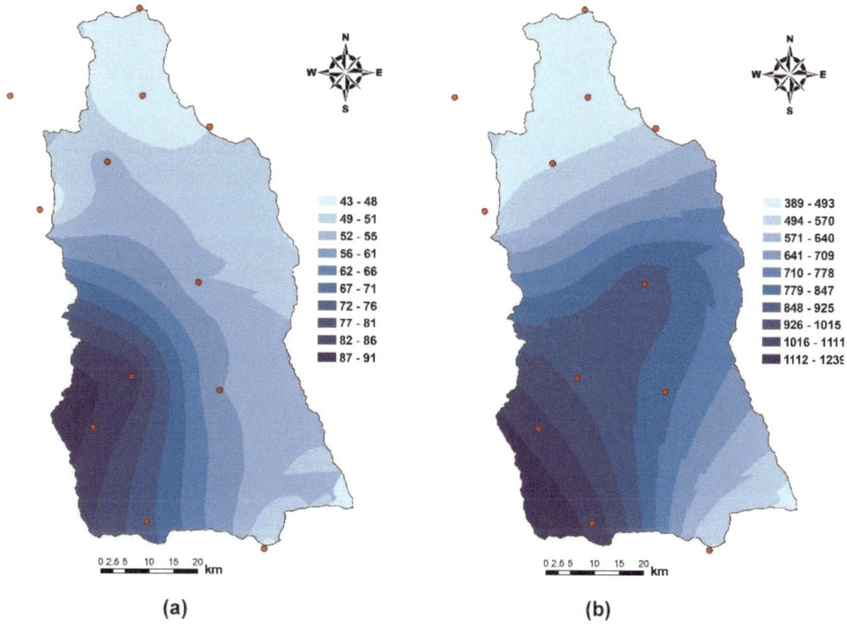

Figure 12.
Rainfall indices of Typhoon Morakot: (a) I_{max} *and (b)* R_{te}.

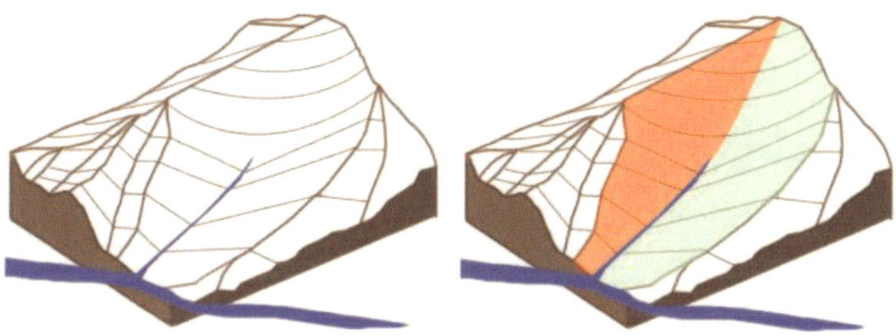

Figure 13.
Slope unit delineation, the left and right slope units of a watershed [3, 4].

Since both the rainfall intensity and rainfall accumulation contribute to the probability of triggering a landslide, the bivariate lognormal distribution was applied in the developing LFC model [4, 14], as in Eq. (3):

$$f_j(x,y) = \frac{1}{2\pi y \zeta_{xj}\zeta_{yj}\sqrt{1-\rho_j^2}} \exp\left\{-\frac{1}{2\left(1-\rho_j^2\right)}\left[\frac{\ln\left(x/c_{xj}\right)^2}{\zeta_{xj}^2} - 2\rho_j\frac{\ln\left(x/c_{xj}\right)\ln\left(y/c_{yj}\right)}{\zeta_{xj}\zeta_{yj}} + \frac{\ln\left(y/c_{yj}\right)^2}{\zeta_{yj}^2}\right]\right\} \tag{3}$$

where $-\infty < x, y, \zeta_{xj}, \zeta_{yj} < \infty$, $c_{xj} > 0$, $c_{yj} > 0$, and $-1 < \rho_j < 1$. In Eq. (3), x and y are maximum hourly rainfall and effective accumulated rainfall, respectively; c_{xj} and c_{yj} are the median; ζ_{xj} and ζ_{yj} are log-standard deviation; ρ_j is the correlation coefficient of x and y. Because the maximum hourly rainfall and the effective accumulated rainfall are treated independently, the ρ_j is zero. Thus, the cumulative density function of Eq. (3) becomes as follows:

$$F_j\left(x, y; c_{xj}, c_{yj}, \zeta_{xj}, \zeta_{yj}\right) = \frac{1}{4}\left[1 + erf\left(\frac{ln\left(x/_{c_{xj}}\right)}{\zeta_{xj}\sqrt{2}}\right)\right]\left[1 + erf\left(\frac{ln\left(y/_{c_{yj}}\right)}{\zeta_{yj}\sqrt{2}}\right)\right] \quad (4)$$

Eq. (4) represents the j-th fragility curve of landslide, including four fragility parameters. The cumulative density function of Eq. (4) is a fragility surface of probability.

The parameters in Eq. (4) can be obtained by using the least square estimate. When the landslide locations and areas are available, meaning the classification of landslide based on the factors (see next section), the fragility curve of landslide (a surface) of a specific classification can be determined.

4.2 Classification of factors

The environmental factors, geology, slope, distance to river, slope aspect, and vegetation index, were classified into levels in order to group similar slope units. The triggering factors of rainfall intensity and effective accumulated rainfall were also redistributed onto slope unit scale. These factors were classified into groups, i.e., two groups of G, three of S, two of R, two of A, and two of N (**Tables 4–8**),

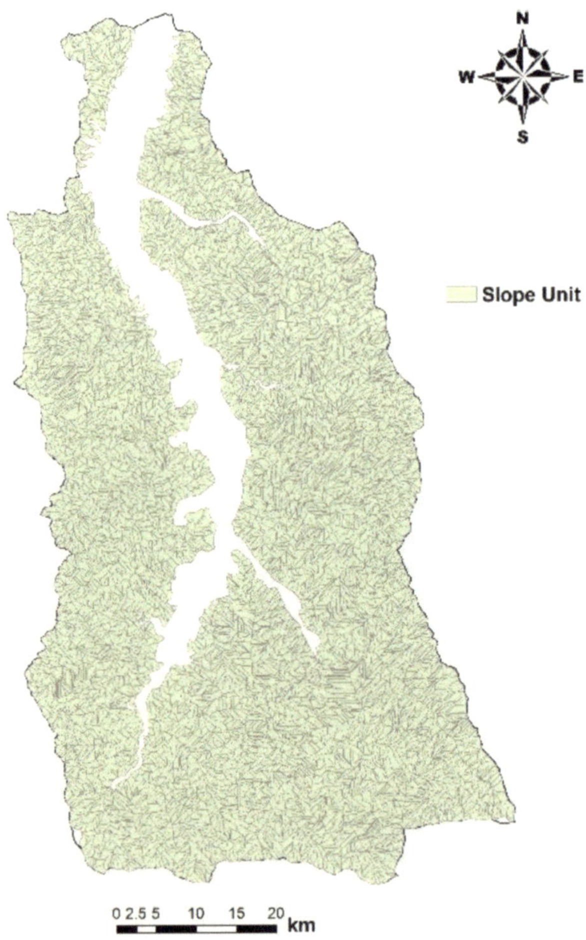

Figure 14.
The slope units of Chen-Yu-Lan watershed.

based on the available data and appropriate judgment to simplify the process. There were total of 48 combinations of classification, as described below.

4.2.1 Geology

The geology is an important factor when considering the potential of landslide. However, the geological conditions, like soil layer depth, rock type, and strength at the site, are not usually available to researchers. Therefore, a simplified step can be used at the geology time scale to generally represent the older and younger stratum of the study area. For Chen-Yu-Lan watershed, the rock type of the area was first used to highlight the geological time scale. The same geology era contained different rock formations, and the factor of geology was classified into two groups, as shown in **Table 4** and **Figure 15**. It was noted that there are 1798 slope units of G1 and 2463 slope units of G2.

4.2.2 Hillside slope

Based on the Soil and Water Conservation Bureau manual, the hillside slope is classified as seven levels. In the fragility model, level 3 to level 7 slopes were considered and simply further classified as three groups, as shown in **Table 5**. **Figure 16** shows the classification results in the Chen-Yu-Lan watershed, and 137 slope units were classified as S1, 827 as S2, and 3297 as S3.

4.2.3 Distance to nearest river channel

The distance to the nearest river channel was classified into two groups, with the threshold value of 300 m. **Table 6** and **Figure 17** show the classification results, in which there are 2482 and 1779 slope units of R1 and R2, respectively.

Classification	Geology time scale	Rock type
G1	Eocene	Dark gray slate and phyllite slate, interbedded with quartz sandstone
	Eocene	Slate and phyllite quartzite sandstone
	Oligocene	Hard shale sandwiched to thick sandstone
	Oligocene	Thick or massive white medium to very coarse quartzite and hard shale
G2	Miocene	Hard shale, slate, phyllite sandstone
	Mid-Miocene	Sandstone and shale interbed, coal seam
	Late Miocene	Sandstone and shale interbed, coal seam
	Miocene to Pliocene	Sandstone and shale interbed, coal seam
	Pliocene	Shale, sandy shale, mudstone
	Pliocene	Sandstone, mudstone, shale interbed
	Pliocene to Pleistocene	Gravel
	Pleistocene	Gravel, sand, and clay

Table 4.
The geology classification.

Classification	SWCB slope level	Technical regulations for soil and water conservation	
		Slope range	degree (°)
S1	3	15% < S ≦ 30%	8.53 < S ≦ 16.70
	4	30% < S ≦ 40%	16.70 < S ≦ 21.80
S2	5	40% < S ≦ 55%	21.80 < S ≦ 28.81
S3	6	55% < S ≦ 100%	28.81 < S ≦ 45.00
	7	S > 100%	S > 45.00

Table 5.
The slope classification.

Classification	Definition	Distance (m)
R1	Close	≤300 m
R2	Not close	>300 m

Table 6.
The classification of distance to river.

Classification	Definition
A1	Weak aspect: the four slope aspects of higher ratio of incremental landslide area. In this study, A1 are E, SE, S, and SW
A2	Strong aspect: the four slope aspects of lower RIL. In this study, A2 are W, NW, N, and NE

Table 7.
The classification of slope aspects.

Image process		Classification	
		Low vegetation	Mid-to-high vegetation
		−1 < NDVI ≦ NDVIc*	NDVIc* < NDVI ≦ 1
Pre-event image	Barren land	N1	N1
	Non-barren land	N1	N2

**NDVIc is the threshold value to classify low and mid-to-high vegetation index. In this study, the NDVIc was −0.35.*

Table 8.
The vegetation classification.

4.2.4 Slope aspects

The slope aspect was considered in the beginning to distinguish the range of frequent landslide on a given mountain slope. There are eight slope aspects (**Figure 18**) used in the study that were grouped into two classes as shown in **Table 7** and **Figure 19**, in which there are 2051 and 2210 slope units of A1 and A2, respectively.

4.2.5 Vegetation index

The land cover status was also an important factor when estimating the landslide potential. The normalized difference vegetation index was used to represent the

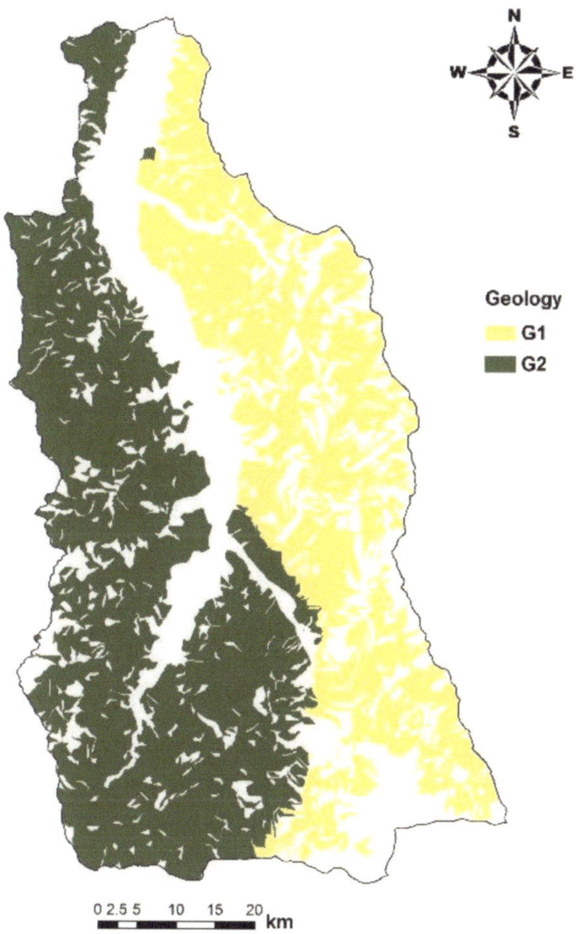

Figure 15.
The geology classification of Chen-Yu-Lan watershed.

land cover status of a given site. Satellite images of SPOT (February 21, 2008, November 28, 2008, and October 14, 2009) were used to calculate the NDVI of the ground surface, and an empirical NDVI threshold was applied to classify barren land and non-barren land. **Table 8** summarized the classification, and **Figure 20** shows the results, in which there are 2765 and 1496 slope units of N1 and N2, respectively.

4.2.6 Maximum rainfall intensity and effective accumulated rainfall

The rainfall data from Typhoon Sinlaku in 2008 and Typhoon Morakot in 2009 was applied to obtain the rainfall intensity and effective accumulated rainfall in the Chen-Yu-Lan watershed. The hourly rainfall data measured at the surrounding weather stations was used to get the rainfall of each slope unit by interpolation. **Figures 21** and **22** show the rainfall distribution during the two typhoon events.

4.2.7 Landslide area

Based on the site investigation in the past after typhoon events, the expected average landslide volume (V) was set as V = 6000 m^3. By applying the relationship of $V = 0.2 \times A^{1.3}$ [15], the landslide area on the slope can be obtained. Therefore, in

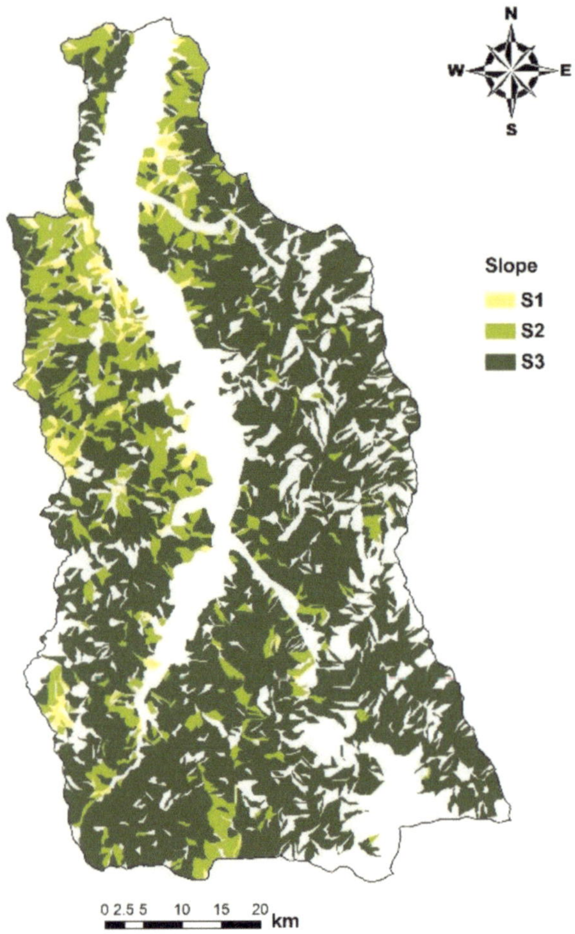

Figure 16.
The slope classification of Chen-Yu-Lan watershed.

association with the slope classification, the determination of landslide of a given slope unit was decided based on the following criteria:

1. Slope S1: the slope unit is counted as a landslide when its landslide area ratio (LAR) is equal to or higher than 5% or the projected landslide area on the slope is greater than 2800 m^2 (0.28 ha). Otherwise, the slope unit is not counted as a landslide area.

2. Slope S2: the slope unit is counted as a landslide when its landslide area ratio is equal to or higher than 5% or the projected landslide area on the slope is greater than 2400 m^2 (0.24 ha). Otherwise, the slope unit is not counted as a landslide area.

3. Slope S3: the slope unit is counted as a landslide when its landslide area ratio is equal to or higher than 5% or the projected landslide area on the slope is greater than 2200 m^2 (0.22 ha). Otherwise, the slope unit is not counted as a landslide area.

The landslide area classification of Chen-Yu-Lan watershed is shown in **Figure 23**. There were 1810 slope units of landslide after Typhon Sinlaku and 1544 ones after Typhoon Morakot, as shown in colored slope units in **Figure 23**.

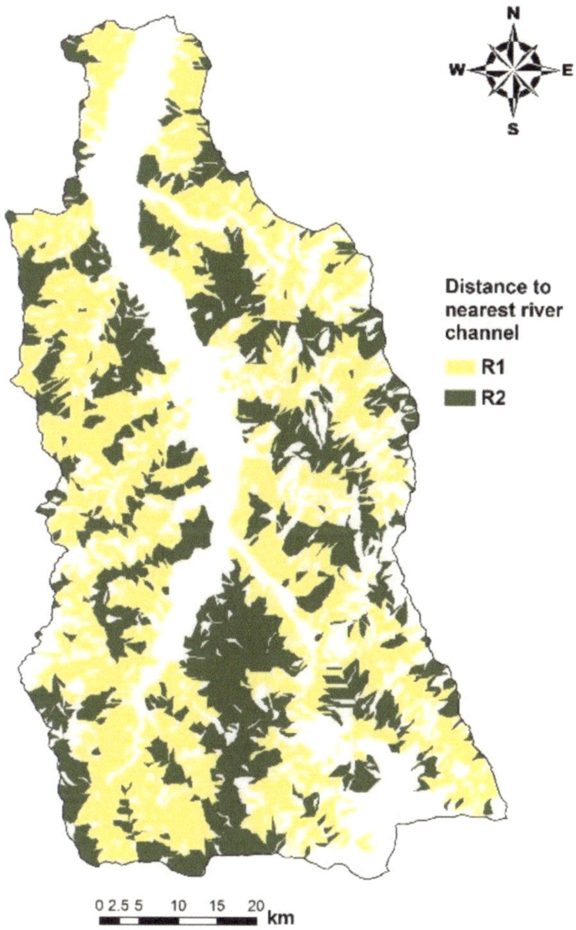

Figure 17.
The classification of distance to the river of Chen-Yu-Lan watershed.

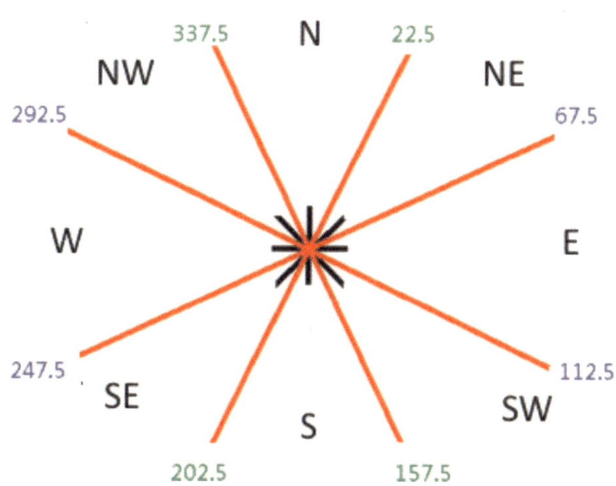

Figure 18.
The slope aspects.

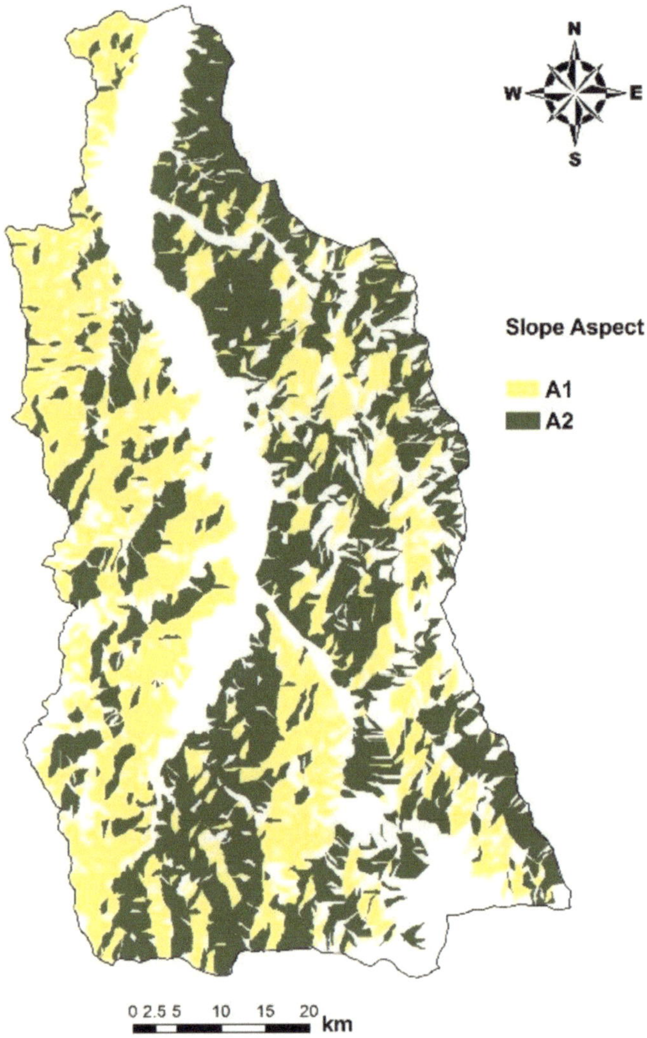

Figure 19.
The slope aspect classification of Chen-Yu-Lan watershed.

4.3 The LFC of Chen-Yu-Lan watershed

The environmental database and rainfall data of typhoon events were applied to classify the slope units and the landslide areas. With the classification described in previous sections, there were a total of 48 classes with combinations of factors G, S, A, R, and N. Each classification was in association with two rainfall indices, the rainfall intensity and effective accumulated rainfall. The fragility of landslide, or the probability of exceeding a level of hazard, was constructed and used for landslide potential assessment. **Tables 9** and **10** summarized the fragility parameters obtained from the two events, and some examples of fragility curves were shown in **Figure 24**. It should be noted that during the classification, insufficient samples of certain classification had led to difficulty of finding parameters needed. Therefore, these samples were combined with other classifications in order to get reasonable probability values of median and standard deviation.

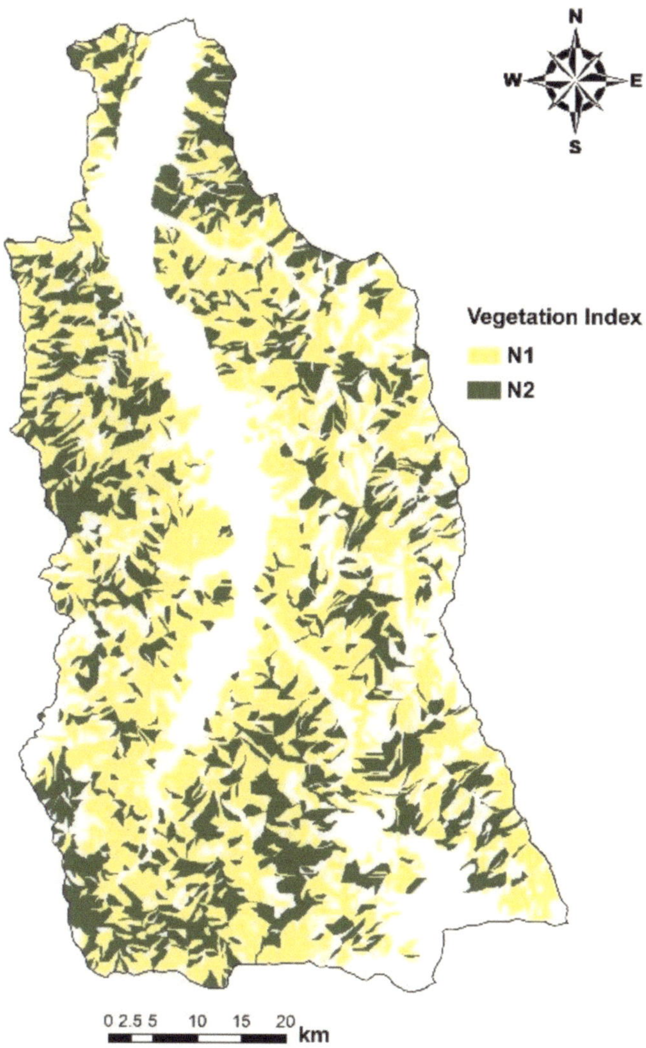

Figure 20.
The vegetation index classification of Chen-Yu-Lan watershed.

4.4 The LFC of Shenmu area

The fragility curves of 48 classification slope units represented the local environmental characteristics of a given area. Instead of directly using 48 set fragility curves, it should be practical to obtain one set of representative fragility curve for a given site or location. To achieve this goal, the weighted fragility curves were introduced and applied to the Shenmu village. The weighted fragility parameters were determined using the following equations:

$$c_{x,m} = \sum_{i=1}^{m} w_i \times c_{xi}, c_{y,m} = \sum_{i=1}^{m} w_i \times c_{yi} \tag{5}$$

$$\zeta_{x,m} = \sqrt{\sum_{i=1}^{m} (w_i \times \zeta_{xi})^2}, \zeta_{y,m} = \sqrt{\sum_{i=1}^{m} \left(w_i \times \zeta_{yi}\right)^2} \tag{6}$$

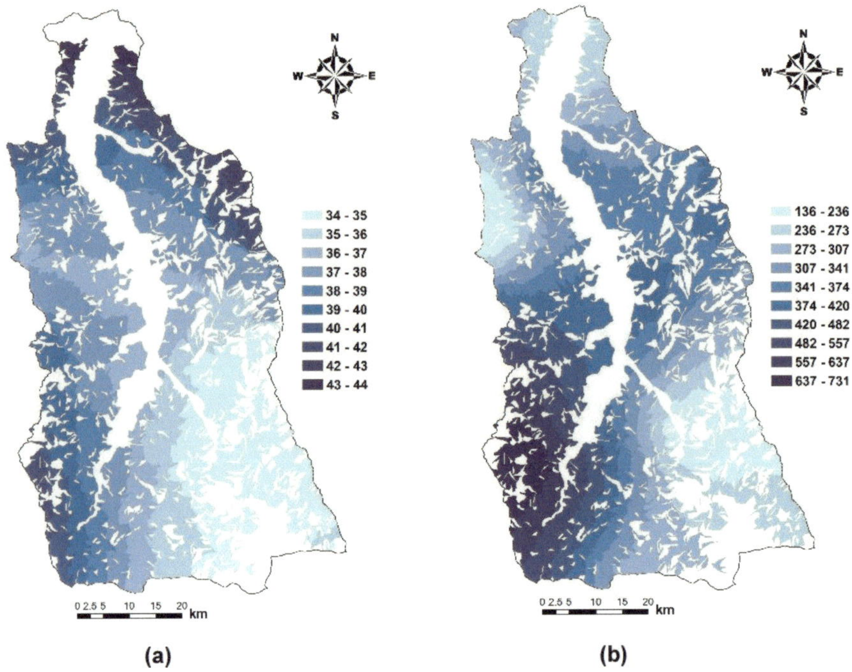

Figure 21.
The rainfall of Chen-Yu-Lan watershed during Typhoon Sinlaku: (a) max. hourly rainfall (I_{max}) and (b) effective accumulated rainfall (R_{te}).

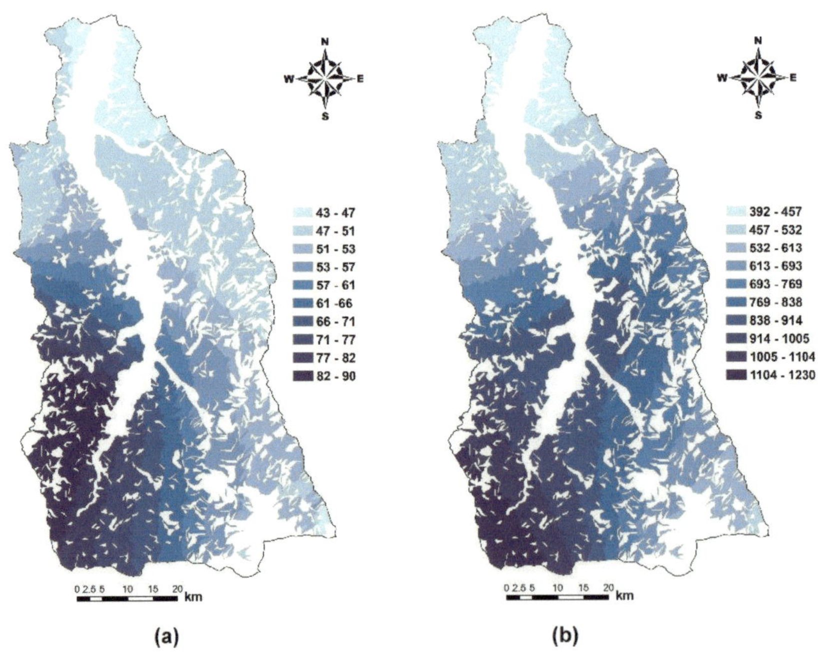

Figure 22.
The rainfall of Chen-Yu-Lan watershed during Typhoon Morakot: (a) max. hourly rainfall (I_{max}) and (b) effective accumulated rainfall (R_{te}).

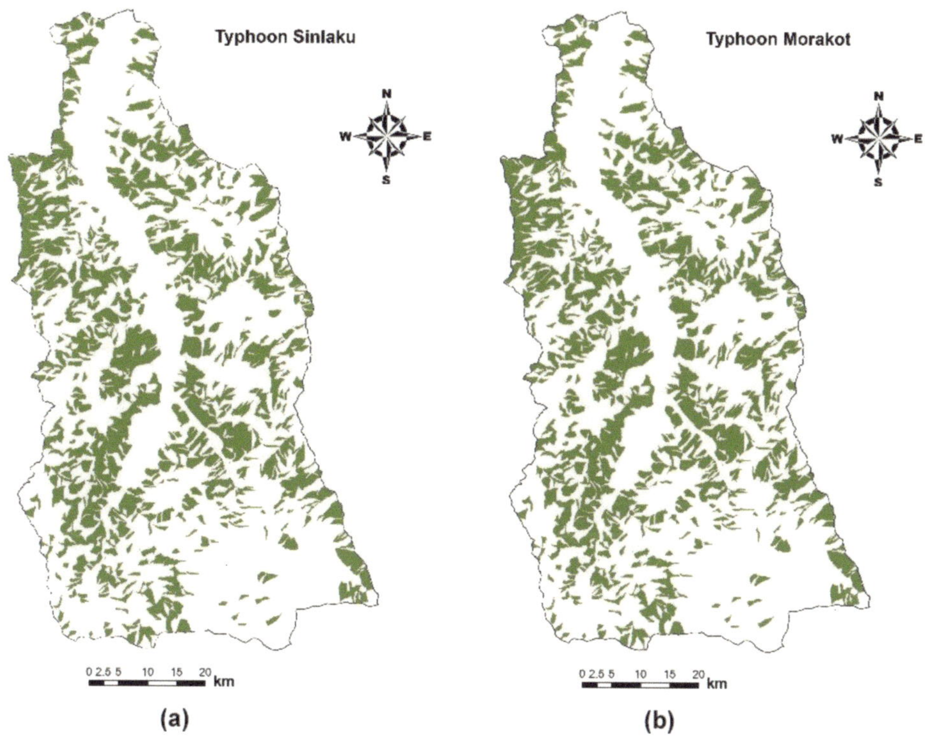

Figure 23.
The landslide area of Chen-Yu-Lan watershed during (a) Typhoon Sinlaku and (b) Typhoon Morakot.

Classification	I_{max} (mm)		R_{te} (mm)		Combined with*
	Median	Std. deviation	Median	Std. deviation	
G1S1A1R1N1	64.40	0.21	485.00	0.28	With 21111
G1S1A1R1N2	27.53	1.24	383.77	0.29	With 21112
G1S1A1R2N1	33.70	0.31	1112.62	0.10	With 21121
G1S1A1R2N2	37.94	0.16	239.39	0.27	With 21122
G1S1A2R1N1	44.40	1.10	290.86	0.24	With 21211
G1S1A2R1N2	43.91	0.16	1007.19	0.71	With 21212
G1S1A2R2N1	32.48	0.77	320.60	0.39	With 21221
G1S1A2R2N2	40.58	0.45	332.07	0.22	With 21222
G1S2A1R1N1	40.44	0.58	235.49	0.79	With 22111
G1S2A1R1N2	72.70	0.32	384.00	0.67	With 22112
G1S2A1R2N1	22.60	0.34	407.35	0.26	With 22121
G1S2A1R2N2	74.16	1.17	527.59	1.20	With 22122
G1S2A2R1N1	22.41	0.70	399.60	1.23	With 22211
G1S2A2R1N2	42.39	0.28	252.25	0.62	With 22212
G1S2A2R2N1	14.08	0.11	706.36	0.80	With 22221
G1S2A2R2N2	115.74	0.61	207.21	0.77	With 22222
G1S3A1R1N1	18.81	0.21	135.69	1.06	

Classification	I_{max} (mm)		R_{te} (mm)		Combined with*
	Median	Std. deviation	Median	Std. deviation	
G1S3A1R1N2	14.51	0.12	295.58	0.29	
G1S3A1R2N1	75.05	0.29	225.74	0.88	
G1S3A1R2N2	28.07	0.38	269.76	0.55	
G1S3A2R1N1	35.79	0.57	967.74	0.35	
G1S3A2R1N2	44.53	1.54	554.12	1.26	
G1S3A2R2N1	29.66	0.72	298.05	0.30	
G1S3A2R2N2	34.00	0.89	269.00	0.69	

**Due to the insufficient data, some classifications were combined together in order to obtain reasonable parameters.*

Table 9.
Fragility parameters of G1 classification.

Classification	I_{max} (mm)		R_{te} (mm)	
	Median	Std. deviation	Median	Std. deviation
G2S1A1R1N1	64.40	0.21	485.00	0.28
G2S1A1R1N2	27.53	1.24	383.77	0.29
G2S1A1R2N1	33.70	0.31	1112.62	0.10
G2S1A1R2N2	37.94	0.16	239.39	0.27
G2S1A2R1N1	44.40	1.10	290.86	0.24
G2S1A2R1N2	43.91	0.16	1007.19	0.71
G2S1A2R2N1	32.48	0.77	320.60	0.39
G2S1A2R2N2	40.58	0.45	332.07	0.22
G2S2A1R1N1	40.44	0.58	235.49	0.79
G2S2A1R1N2	72.70	0.32	384.00	0.67
G2S2A1R2N1	22.60	0.34	407.35	0.26
G2S2A1R2N2	74.16	1.17	527.59	1.20
G2S2A2R1N1	22.41	0.70	399.60	1.23
G2S2A2R1N2	42.39	0.28	252.25	0.62
G2S2A2R2N1	14.08	0.11	706.36	0.80
G2S2A2R2N2	115.74	0.61	207.21	0.77
G2S3A1R1N1	16.70	0.13	604.42	0.53
G2S3A1R1N2	72.54	0.58	305.93	0.41
G2S3A1R2N1	21.81	1.31	387.14	0.84
G2S3A1R2N2	56.01	1.07	527.88	0.69
G2S3A2R1N1	23.20	0.78	378.00	0.66
G2S3A2R1N2	14.50	0.11	151.30	0.10
G2S3A2R2N1	23.76	0.66	270.92	0.28
G2S3A2R2N2	29.86	1.02	249.28	0.80

Table 10.
Fragility parameters of G2 classification.

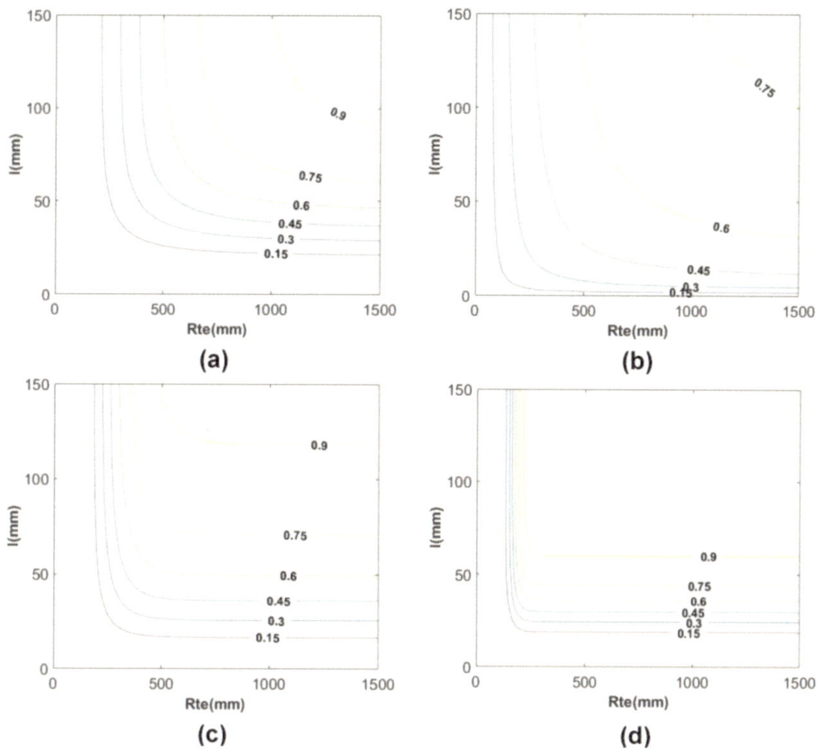

Figure 24.
Examples of fragility curves of Chen-Yu-Lan watershed: (a) G1S3A1R1N1, (b) G2S2A1R1N1, (c) G1S3A1R2N1, and (d) G2S3A1R2N1.

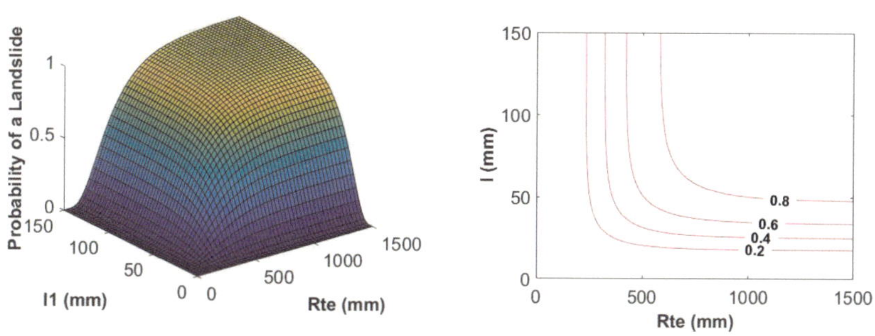

Figure 25.
The fragility surface and fragility curves of Shenmu area.

$$w_i = \frac{n_i}{N_i} \tag{7}$$

where x and y are rainfall indices, $c_{x,\,m}$ and $c_{y,\,m}$ are the weighted median values, $\zeta_{x,\,m}$ and $\zeta_{y,\,m}$ are weighted standard deviation, m is the number of classifications, w_i is the weighting factor of a classification, n_i is the number of slope units of a given classification, and N_i is the total number of slope units.

After the weighted calculation, the fragility parameters of Shenmu area are median $I_{max} = 33$ mm and median $R_{te} = 413$ mm. **Figure 25** shows the weighted fragility curves of Shenmu area.

5. Case studies and results

The risk of landslide was demonstrated by using the critical values of rainfall hazard and landslide fragility. The concept of landslide warning was adopted in this study, and by combining both *Hc* and *Fc*, the warning status includes safe stage and unsafe stages, as illustrated in **Figure 26**. It should be noted that there are two stages of unsafe status, Red I and Red II. Red I stage indicates that the situation has pass *Hc* and a rainfall hazard could occur. Red II stage implies the most serious condition that in addition to the rainfall hazard, a landslide could occur as well. Both stages are determined with a probability when given a rainfall condition. The procedure of determining safe stage was designed to match the needs of disaster preparation and prediction of government.

Cases of landslides and debris flows in Shenmu were collected from the disaster notices issued by Soil and Water Conservation Bureau of Taiwan. As shown in **Table 11** and **Figure 27**, a total of seven cases were used to determine the critical values of *Hc* (=0.91) and *Fc* (=0.23) of Shenmu. These cases were used in the assumption that whenever there was a debris flow, there should be landslides at the upper stream areas before or during the debris flow.

The rainfall history of Typhoon Morakot in 2009 and 0601 Heavy Rainfall in 2016 were used to evaluate the landslide risk assessment in Shenmu. **Figure 28**

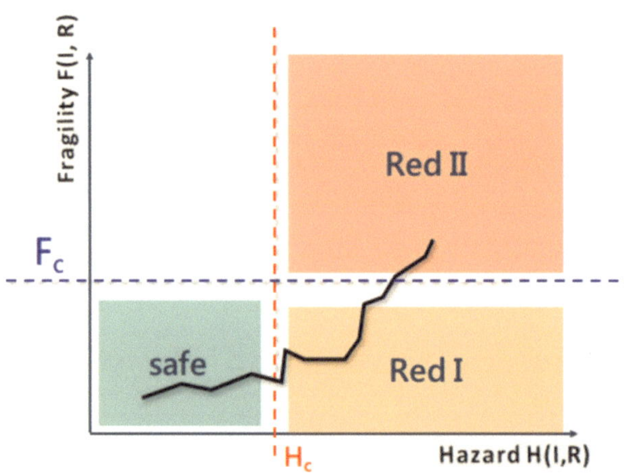

Figure 26.
The warning conditions based on landslide fragility (Fc) and rainfall hazard (Hc).

Year	Event	Disaster	Village	I_{max} (mm)	R_{te} (mm)
2009	Typhoon Morakot	Debris flow, flood	Tongfu	85.5	1130
2009	Typhoon Morakot	Debris flow	Wangmei	85.5	1130
2009	Typhoon Morakot	Landslide	Shenmu	47.5	829.5
2009	Typhoon Morakot	Debris flow	Shenmu	42.5	750
2009	Typhoon Morakot	Debris flow	Shenmu	33.5	641
2009	Typhoon Morakot	Landslide	Shenmu	20	476.5
2009	Typhoon Morakot	Debris flow	Shenmu	38.5	877
2012	0610 Heavy rainfall	Debris flow, flood	Shenmu	18.5	450.6

Table 11.
The disaster notices around Shenmu area.

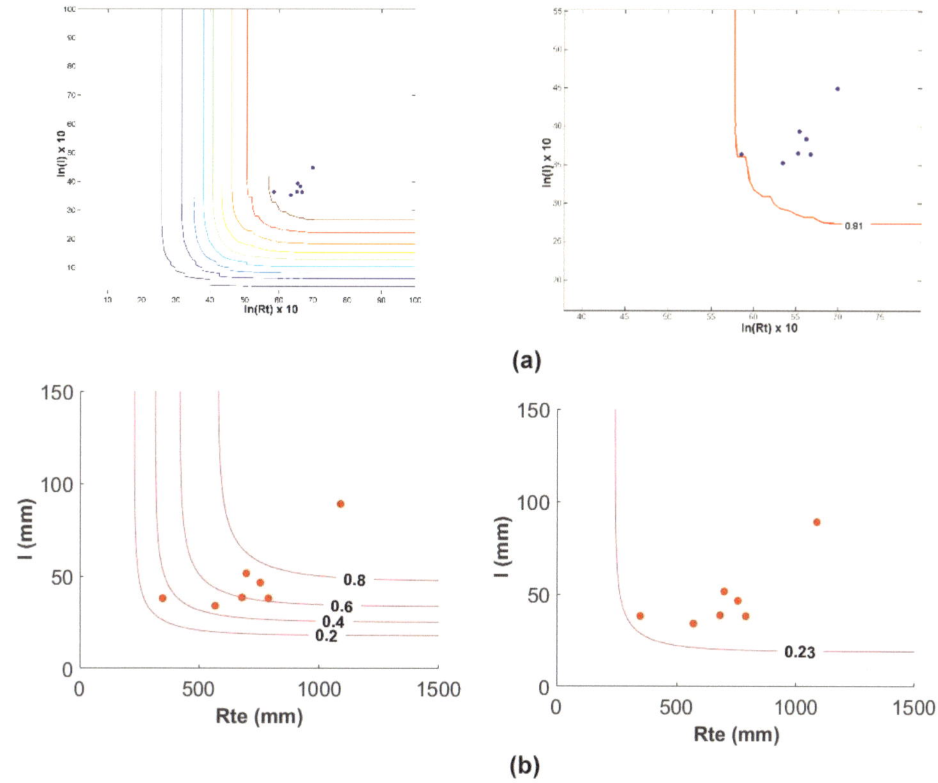

Figure 27.
The probability thresholds of rainfall hazard and landslide fragility in Shenmu area: (a) rainfall warning threshold and (b) landslide warning threshold.

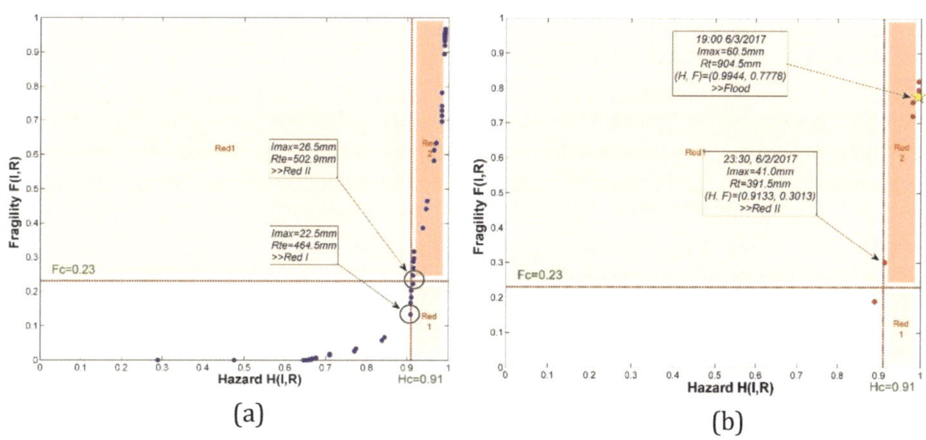

Figure 28.
The change of probability in Shenmu area during (a) Typhoon Morakot (2009) event and (b) 0601 heavy rainfall in 2016 (after [4, 5]).

shows the results of event, and the dots in the figure represent the rainfall condition (hourly rainfall and cumulative rainfall) and the probability of hazard. It was noted that the dots behaved like a "snake" line going from Safe stage to Red I and Red II stages. Also, the snake line stayed shortly at Red I stage for both events and passed to Red II in a jump. This condition implied that when the situation was beyond the *Hc* line, the landslide hazard was very likely to occur. The results conformed to the records of Typhoon Morakot. Severe landslides occurred at the upper stream areas in Shenmu during the typhoon. Therefore, the proposed risk assessment and warning stages of landslide were reasonably useful in this case.

6. Summary and conclusions

This study had developed the landslide fragility curve model by using the spatial data and statistical methods. The fragility curves of the study area were derived for all combinations of environmental and triggering factors. The data sets included the geomorphological and vegetation condition factors, based on the landslides at the Chen-Yu-Lan watershed in Taiwan, during Typhoon Sinlaku (September 2008) and Typhoon Morakot (August 2009). This study also proposed landslide risk assessment using rainfall hazard potential and landslide fragility curves and concluded findings as follows:

1. Overall, the proposed model provides considerably accurate and reliable results on landslide estimations in terms of spatial distribution.

2. Adoption of slope unit was physically proper in modeling landslide locations.

3. The classifications of slope unit can be applied to different areas, and the fragility curve of each classification can be used directly.

4. The procedure of risk assessment was useful for practical landslide disaster preparation and prediction.

5. The LFC model was developed using two typhoon events. More events and landslide cases are needed to improve the LFC model in the future. Furthermore, the classification of upstream areas based on their environment is suggested for better possible estimation.

6. The applicability of factors should be considered before developing the model. The concerns about the model factors and the limits of satellite images can be resolved by using different methods to obtain necessary data. For example, the information of LIDAR may be used with the satellite images to provide better description on landslide identification. Therefore, the LFC model could be improved when more factors are available and applicable.

Acknowledgements

The authors would like to express their gratitude to research assistant Xingping Wang, for helping in collecting all the data relevant to the landslides in the Chen-Yu-Lan watershed. The authors also would like to thank the Soil and Water Conservation Bureau in Taiwan for supporting this research.

Author details

Yi-Min Huang[1*], Tsu-Chiang Lei[2], Bing-Jean Lee[1] and Meng-Hsun Hsieh[1]

1 Department of Civil Engineering, Feng Chia University, Taichung, Taiwan

2 Department of Urban Planning and Spatial Information, Feng Chia University, Taichung, Taiwan

*Address all correspondence to: ninerh@mail.fcu.edu.tw

References

[1] Dilley M, Chen RS, Deichmann U, Lerner-Lam AL, Arnold M, Agwe J, et al. Natural Disaster Hotspots: A Global Risk Analysis (English). Washington, DC: World Bank. Available from: http://documents.worldbank.org/curated/en/621711468175150317/Natural-disaster-hotspots-A-global-risk-analysis; 2005

[2] Lei TC, Huang YM, Lee BJ, Hsieh MH, Lin KT. Development of an empirical model for rainfall-induced hillside vulnerability assessment: A case study on Chen-Yu-Lan Watershed, Nantou, Taiwan. Natural Hazards. 2014; **74**:341-373

[3] Lee BJ, Lei TC, Huang YM, Hsieh MH. 2015 Application of Landslide Fragility Curves in Landslide Risk and Warning Criteria, Project Report. Soil and Water Conservation Bureau, Taiwan (in Chinese). 2016. p. 341

[4] Lee BJ, Lei TC, Huang YM, Hsieh MH. 2016 Application of Landslide Fragility Curves in Landslide Risk and Warning Criteria, Project Report. Soil and Water Conservation Bureau, Taiwan (in Chinese). 2017. p. 224

[5] Lee CY et al. Risk assessment of landslide by using fragility curves: A case study in Shenmu, Taiwan. In: Proceedings of the 5th International Conference on Geotechnical Engineering for Disaster Mitigation and Rehabilitation (5th GEDMAR); 14–17 December 2005; Taipei, Taiwan: Airiti Press Inc. 2017. pp. 137-148

[6] Pradhan B, Lee S. Delineation of landslide hazard areas on Penang Island, Malaysia, by using frequency ratio, logistic regression, and artificial neural network models. Environmental Erath Science. 2009;**60**(5):1037-1054

[7] Lei TC, Wan S, Chou TY, Pai HC. The knowledge expression on debris flow potential analysis through PCA +LDA and rough sets theory: A case study of Chen-Yu-Lan Watershed, Nantou, Taiwan. Environmental Earth Sciences. 2011;**63**(5):981-997

[8] Wan S, Lei TC, Chou TY. A landslide expert system: Image classification through integration of data mining approaches for multi-category analysis. International Journal of Geographical Information Science. 2012;**26**:747-770

[9] Soil and Water Conservation Bureau. Soil and Water Conservation Handbook. Taiwan: Soil and Water Conservation Bureau; 2016. Available from: https://drive.google.com/file/d/1FzwMzkd_12qzGhOGjcQX7F-x2ihHvr6e/view

[10] Lillesand TM, Kiefer RW. Remote Sensing and Image Interpretation. New York: John Wiley and Sons; 1999. 736 p

[11] Bannari A, Morin D, Bonn F, Huete AR. A review of vegetation indices. Remote Sensing Reviews. 1995; **13**:95-120

[12] Xie M, Esaki T, Zhou G. GIS-based probabilistic mapping of landslide hazard using a three-dimensional deterministic model. Natural Hazards. 2004;**33**:265-282. DOI: 10.1023/B:NHAZ.0000037036.01850.0d

[13] Shinozuka M, Feng MQ, Lee J, Naganuma T. Statistical analysis of fragility curves. ASCE Journal of Engineering Mechanics. 2000;**126**(12):1224-1231

[14] Hsieh MH, Lee BJ, Lei TC, Lin JY. Development of medium- and low-rise reinforced concrete building fragility curves based on Chi-Chi Earthquake data. Natural Hazards.2013;**69**(1):695-728. DOI: 10.1007/s11069-013-0733-8

[15] AECOM. A study of sediment management policies on climate change for river basins in southern

Taiwan- Gaoping river case study. Project Report. Water Resources Planning Institute, Water Resources Agency, Ministry of Economic Affairs, Taiwan. 2011. 342 p. Available from: https://www.wrap.gov.tw/library_1.aspx?id=22547 Accessed: 19 July 2019]